THE COMPLETE GUIDE TO SINGLE-ENGINE MOONEYS

Other TAB books by the author:

No. 1147 *How to Build Adobe Houses...etc.*
No. 2262 *How the Air Traffic Control Systems Works*
No. 2263 *A Funny Thing Happened On The Way To The Airport*
No. 2266 *Night Flying in Single-Engine Airplanes*
No. 2274 *The Illustrated Encyclopedia of General Aviation*
No. 2282 *Flying VFR in Marginal Weather*
No. 2284 *Cross-Country Flying*
No. 2286 *Practical Area Navigation*
No. 2296 *The Corporate Aircraft Owner's Handbook*

Dedication

This book is dedicated to Al Mooney—he knew how to design a better airplane—and to his brother Art, who managed to make sure that Al's paper airplanes were turned into flying reality.

I want to express my appreciation for the generous amount of time give to me by Al and Art Mooney, George Haddaway, Norman Hoffman, Ralph Harmon, Jack Womack, Bob Pyle and Roy Lopresti during the researching of this book.

The photographs of Al Mooney's various aircraft reproduced in the chapter on Al and Art Mooney are from Al Mooney's personal collection and used with his permission. All other photographs not otherwise credited are by the author.

THE COMPLETE GUIDE TO SINGLE-ENGINE MOONEYS

by Paul Garrison

MODERN AVIATION SERIES

TAB BOOKS

BLUE RIDGE SUMMIT, PA. 17214

FIRST EDITION

FIRST PRINTING—AUGUST 1980

Printed in the United States of America

Library of Congress Cataloging in Publication Data

Garrison, Paul.
The complete guide to single-engine Mooneys.

Includes index.
1. Mooney airplanes. 2. Mooney, Al. 3. Mooney Aircraft Corporation. 4. Air craft industry—United States—Biography. I. Title.
TL686.M56G37 629.133'343 80-20411
ISBN 0-8306-9698-9
ISBN 0-8306-2283-7 (pbk.)

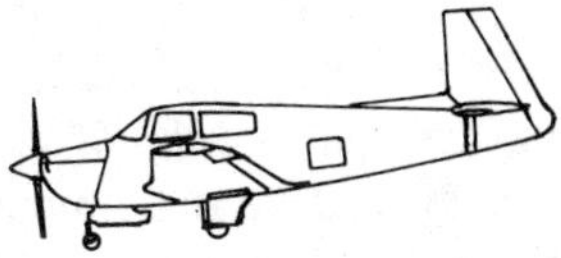

Preface

The Mooney story is, in fact, not one story but three. First and above all, it is the story of an airplane. But it is also the story of three of aviation's great men. And then it is the story of a company, actually a series of companies, which used the name and one basic design to produce a family of high-performance, single-engine airplanes which, for decades, have proven to be the most efficient in their class.

In this book, the basic design for the 4-place Mark 20 is traced through its various growth versions to today's unsurpassed leader in the field—the Mooney Turbo 231. Of the original wood-wing version, a fair number are still flying today. Other topics covered include the conversion to metal wings, the stretched version which added leg room and comfort, the brief departure from the original design (involving T-tails), the somewhat less than successful attempt to manufacture and market the first-ever pressurized single-engine airplane, the painstaking cleanup of the original design which resulted in a 22-mph increase in cruise, and finally, the turbocharged version which added another 30 mph at optimum-altitude cruise.

We will show comparisons of the various models with competitive aircraft of their day, examine each in terms of cost, speed, fuel consumption, and overall efficiency, and there are interviews with some of the people who, from a technical standpoint, were most actively involved in the development of the different growth versions of the original design.

Al Mooney designed and built his first successful airplane in 1925 at age 19 and spent his life from that day, until his recent retirement, in drafty hangars and aircraft factories creating a total of 23 different airplanes that ranged from the original fabric-covered Eaglerock biplane to the jet-powered VTOL Hummingbird. We will follow his successes and frustrations through the 50 years of his active involvement in aviation and we'll show you whatever pictures are still available of the many Mooney-designed aircraft.

We will then meet the two men who used AL Mooney's original design and gradually changed and improved it. Ralph Harmon redesigned the original Mark 20, with its wooden wings and tail, to produce an all-metal aircraft and he was also the designer of the ill-fated pressurized Mooney Mustang. And there is Roy Lopresti who is responsible for turning the Mooney Executive into the 201 and 231. We'll also take a look at the company, or rather the series of companies, which manufactured and marketed the airplane which we now know as *The Mooney*. That airplane was originally certificated as the M-20, Al Mooney's 20th design, in July of 1954. We'll cover the involvement with other companies, the assembly and marketing agreement with Mitsubishi Heavy Industries, the financial problems which resulted in the eventual sale of the company and the subsequent bankruptcy, and finally the rebirth of the Mooney Aircraft Corp. as an all-new company—a subsidiary of Republic Steel Corp.

Along the way, there are interviews with not only Al Mooney and his brother Art, but also with others who played a key role: George Haddaway, who was largely responsible for Al's move from Wichita, Kansas, to Kerrville, Texas; Norm Hoffman, who was in charge of marketing and who singlehandedly turned a totally unknown airplane into the best-selling retractable single of its day; Jack Womack whose advertising helped the spread the word; Charlie Dugosh, who knows more about maintenance and how to keep a Mooney flying than anyone else in the world; and Roy Lopresti, who designed and supervised the cleanup which brought us the 201 and the Turbo 231 and who is now in the throes of designing what in the company is known as the M-X, a 6-place pressurized single with a design cruising speed of 300 mph.

Paul Garrison

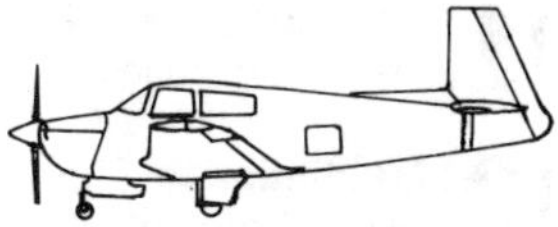

Contents

Chapter 1

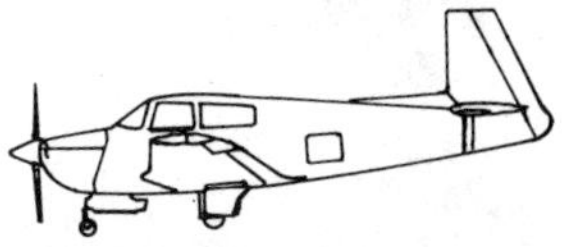

The Mooney Syndrome

Throughout the company's long and occasionally checkered career, Mooney airplanes have always had their fair share of detractors as well as an ample number of those who loved them with a passion. The biggest complaint the detractors would voice with depressing regularity was that the cabin is simply too small for comfort, that occupants of the rear seats better be amputees because there is simply no room for legs, and that the performance figures in terms of speed and economy were greatly exaggerated.

Early Models

About the early Mooneys (Fig. 1-1) with the manual gear-retraction system, they also liked to say that retracting or lowering the gear required a degree of dexterity which could only be expected from a contortionist. When PC (positive control) was introduced as standard equipment on all Mooneys, some purists among pilots (Richard Bach, author of *Jonathan Livingston Seagull*, for one) complained that it removed control from the pilot. Some were known to place a Band-Aid permanently over the cut-off button on the yoke to disable its wing-leveling action. And then there were the nitpickers who felt that displaying the fuel quantity in pounds rather than gallons was a bad feature.

Claims

The Mooney lovers, on the other hand, would claim that their birds flew as fast or faster than the book said they should. They loved their manual gear which could be retracted in a fraction of a second. They enjoyed the feeling of, for all practical purposes, "wearing" the airplane which resulted from the fact that the pilot's legs are buried deep under the instrument panel. And they felt that reading the fuel quantities in pounds produced greater accuracy than would have been possible with gallons. And most pilots, especially those who used their airplanes habitually for long cross-country flights, liked the PC (a Brittain wing-leveler) which made taking the hands and feet off the controls for prolonged periods of time easy and safe.

Facts

There are truths, half-truths and untruths in many of these statements. True, the cabin is somewhat smaller than that of competing aircraft, but the difference, when actually measured with a tap measure, is so slight as to be meaningless. It's barely an inch or two narrower and rearseat passengers in, for example, an Arrow have actually less legroom than those in the later stretched Mooney models (the Statesman, the Executive, the 201 and 231). The cabin height is limited and pilots who like to strut around in a top hat or an overgrown Stetson will find that they have to take their hat off.

Mooneys have always been fast, but not faster than the book claims, and there was a time when the book figures were considerably exaggerated in order to make the airplane seem faster than it actually was. But regardless of the validity of the manufacturer's claims for performance parameters, on a dollar-for-dollar basis, every single Mooney model has always been the most efficient and economical in its class.

The manual gear could be retracted or extended in a fraction of a second, but it did require some rather awkward contortions and a fair degree of practice. Most of the time everything would work just fine. But if the pilot, for one reason or another, happened to fail to complete the extension or retraction in one continuous coordinated movement, it would be a difficult job to get the gear either up or down because the airstream somehow seemed to try to keep it in the halfway position.

You do have to get used to the nearness of the instrument panel once the pilot's seat has been moved sufficiently far forward to permit reaching the rudder pedals. In fact, first-time Mooney pilots tend to get the feeling that their legs are buried somewhere in the engine compartment.

Fig. 1-1. Al Mooney's original Mark 20 (Courtesy of Al Mooney).

Features

The decision to provide an always-on wing leveler as standard equipment, while it may have offended some, in retrospect seems to have been a good idea. Any pilot who found himself inadvertantly in IFR conditions or in those haze conditions in which reference to the horizon becomes tenuous, if not nonexistent, was never in any danger of finding himself in an unusual attitude and the related possibility of over stressing the airplane during a pullout. It might, of course, have been responsible for some VFR pilots flying inten-

Fig. 1-2. The cabin of the Mooney is built around a welded steel-tube cage (courtesy of Mooney).

tionally into the clouds, the fear of not being able to control the airplane by reference to instruments having been removed, but this is strictly speculation.

One of the Mooney features about which there is generally complete agreement is the soundness of its basic structure. The airplane started out as a wooden airplane built around a steel-tube cage and an exceedingly strong continuous wing spar. These two features were retained when it became an all-metal airplane. The welded steel-tube cage (Fig. 1-2), much like a collection of multi-directional roll-bars, provides the cabin with a degree of crash-worthiness which exceeds all other airplanes in its class and which has been known to be responsible for saving a considerable number of lives. It is still the basic structure of the 201s and 231s (Fig. 1-3) today. And the wing spar (Fig. 1-4), one continuous structure which goes through the cabin, is unbelievably strong. There are authenticated cases of Mooneys having been abruptly pulled out of a 300-mph dive, causing the wings to bend to a point that hundreds of rivets would pop, but subsequent examinations would show that the wing was not deformed and had suffered no actual structural damage.

Fig. 1-3. Today's leader among high-performance, single-engine airplanes—the Mooney Turbo 231.

Mooneys are tough airplanes which provide the pilot and passengers with a safe environment if the pilot has the experience and proficiency to handle a fast, clean high-performance machine.

A pilot advancing from, for example, a Cessna Skyhawk or Skylane to a Mooney must expect that it will take a little while to get used to the peculiarities and flying characteristics of a Mooney. First of all, the airplane has short legs. It is built close to the ground and, as a result, it is particularly susceptible to ground effect. It tends to want to fly long before actual flying speed has been reached. Experienced Mooney pilots have been known to habitually lift it off the ground just a few inches, then pull up the gear and continue in level flight until sufficient speed for climb has been reached (not a recommended or very safe procedure).

During climb and cruise, there is no appreciable difference to be noted. However, in high-altitude cruise above haze and dust conditions, the ram-air feature is unique. It permits a degree of extra power by bypassing the air filter. It is the descent which requires special attention. The Mooney is a very clean airplane and slowing it down to an acceptable approach speed is not always easy. It seems to simply want to keep on going. Today, in the new

Fig. 1-4. The Famous continuous wing spar is unbelievably strong.

models, the gear-lowering speed has been upped to 152 mph (IAS)(132 knots), thus providing a useful speed brake. But on older models, the gear-extension speed was 120 mph (104 knots), and the only way to get the airplane to slow down was to reduce power and pull the noise way up. Since this would usually result in an initial gain in altitude, it was (and still is) important to initiate the slowdown routine a sufficient distance from the destination airport to be able to arrive at pattern altitude with a minimum of excess speed. And, once down near the ground, the airplane hates to sit down. It likes to float inches above the runway in ground effect. Many pilots routinely will then simply retract the flaps in order to force it down onto the concrete.

Stalls

There has been talk of Mooneys having a tendency to get into a flat spin from which recovery is extremely tricky—if not impossible. In many hundreds of hours of flying Mooneys, including all manner of power-on and power-off practice stalls, I personally have never experienced this tendency. I don't believe that it is something to worry about when the airplane is flown more or less the way it is supposed to be flown.

It is not intended to be an aerobatic airplane. While pilots have been known to roll and loop them, they do so at their own risk. The purchaser of a Mooney once asked Bob Pyle (the dealer who had sold it to him) during a familiarization flight whether the PC would be capable of righting the airplane from an inverted position or of keeping it level while inverted. Bob said he didn't know but that he was willing to try. They climbed to 12,000 feet and he rolled the airplane on its back, then took his hands and feet off the controls to see what it would do. It went into a screaming dive and eventually required some help from the pilot to get back to straight-and-level flight before too much altitude was lost. Don't try it! Another Mooney dealer liked to demonstrate to prospective customers the outstanding glide capability of the airplane. With the home airport miles away and still out of sight, he would shut down the engine and then casually glide home—always making a perfect dead-stick landing. It goes without saying that he knew the exact spot at which to shut down the engine while assuming no major adverse wind conditions.

Wood-Wings

What about those old wood-wing Mooneys that are still flying and are occasionally offered for sale? While there is nothing wrong

Table 1-1. The M-20 Family of Aircraft.

Year	Model	Name	Gross weight	hp	Engine	Maximum speed	Cruise speed	Price	Remarks
1953	M-20	Mark 20	2200	145	C/C-145	?	?	NA	Wooden wing & tail
1955	M-20	Mark 20	2450	150	L/O-320	171	165	12,500	
1957	M-20	Mark 20	2450	150	L/O-320	171	165	13,750	
1958	M-20A	Mark 20A	2450	180	L/O-360-A1A	190	180	14,750	
1960	M-20A	Mark 20A	2450	180	L/O-360A1A	190	180	15,450	
1961	M-20B	Mark 21	2450	180	L/O-360-A1A	190	182	15,995	All metal
1962	M-20C	Mark 21	2575	180	L/O-360-A1D	190	182	15,995	
1963	M-20C	Mark 21	2575	180	L/O-360-A1D	190	182	16,450	
1963	M-20D	Master	2500	180	L/O-360-A1D	147	135	13,995	Fixed gear, fixed-pitch prop
1964	M-20E	Super 21	2575	200	L/IO-360-A1A	197	187	18,450	
1965	M-20C	Mark 21	2575	180	L/O-360-A1D	190	182	16,450	PC standard; electric gear optional
1965	M-20E	Super 21	2575	200	L/IO-360-A1A	197	187	18,450	PC standard; electric gear optional
1966	M-20C	Mark 21	2575	180	L/O-360-A1D	190	182	16,950	
1967	M-20C	Mark 21	2575	180	L/O-360-A1D	190	182	18,250	
1967	M-20E	Super 21	2575	200	L/IO-360-A1A	197	187	19,995	
1967	M-20F	Executive 21	2740	200	L/IO-360-A1A	184	197	21,995	Stretched 10 inches
1968	M-20C	Ranger	2575	180	L/O-360-A1D	179	172	16,600	
1968	M-20G	Statesman	2525	180	L/O-360-A1D	178	169	18,790	Stretched version
1969	M-20C	Ranger	2575	180	L/O-360-A1D	179	172	17,480	
1969	M-20G	Statesman	2525	180	L/O-360-A1D	178	169	20,400	
1969	M-20E	Chaparral	2575	200	L/IO-360-A1A	190	182	20,755	
1969	M-20F	Executive	2740	200	L/IO-360-A1A	184	179	23,200	
1970	M-20C	200 Ranger	2575	180	L/O-360-A1D	179	172	18,995	Butler-Aerostar/Mooney
1970	M-20E	201 Chaparral	2575	200	L/IO-360-A1A	190	182	21,930	Electric gear standard
1970	M-20F	220 Executive	2740	200	L/IO-360-A1A	184	179	23,852	
1971	M-20C	200 Aerostar	2575	180	L/O-360-A1D	179	172	19,995	
1971	M-20E	201 Aerostar	2575	200	L/IO-360-A1A	190	182	21,995	
1971	M-20F	220 Aerostar	2740	200	L/IO-360-A1A	184	179	23,995	
1974	M-20C	Ranger	2575	180	L/O-360-A1D	18C	170	26,500	Subsidiary of Republic Steel
1974	M-20E	Chaparral	2575	200	L/IO-360-A1A	185	175	28,500	
1974	M-20F	Executive	2740	200	L/IO-360-A1A	180	170	30,500	
1975	M-20C	Ranger	2575	180	L-O-360-A1D	180	170	28,720	
1975	M-20E	Chaparral	2575	200	L-IO-360-A1A	185	175	30,515	
1975	M-20F	Executive	2740	200	L-IO-360-A1A	180	170	33,065	
1976	M-20C	Ranger	2575	180	L/O-360-A1D	180	170	35,000	
1976	M-20E	Chaparral	2575	200	L/IO-360-A1A	185	175	37,258	
1976	M-20F	Executive	2740	200	L-IO-360-A1A	180	170	39,785	
1977	M-20C	Ranger	2575	180	L/O-360-A1D	180	170	30,200	
1977	M-20F	Executive	2740	200	L/IO-360-A1A	180	170	34,850	
1977	M-20J	201	2740	200	L/IO-360-A1B6D	201	182	41,200	
1978	M-20C	Ranger	2575	180	L/O-360-A1D	180	170	32,700	
1978	M-20J	201	2740	200	L/IO-360-A1B6D	201	182	43,500	
1979	M-20J	201	2740	200	L/IO-360-A3B6D	225	200	46,725	
1979	M-20K	231	2900	210	C/TSIO-360-GB	225	200	51,975	Turbocharged

with a perfectly manufactured and well-maintained wooden wing, wood does tend to deteriorate with age. With hardly anyone left today who knows how to repair wooden wings—the only one I know of is Charley Dugosh on the Kerrville airport—these old airplanes must be approached with great caution. Wood-wing airplanes that were tied down rather than hangared for long periods of time are likely to show areas of deterioration which will be costly to repair. Similarly, such aircraft which were based in areas with either extremely damp or extremely dry climates might show detrimental effects of those climates in the wood. There has been a larger-than-normal number of structural failures of these airplanes in recent years. Obviously, this is the result of increasing age and insufficient maintenance.

Still, a well-maintained wood-wing airplane is lighter than its metal counterpart and it is aerodynamically cleaner than any of the metal Mooneys produced with the exception of the 201 and 231. But insurance might be somewhat difficult to come by and, when available, it's bound to be expensive.

New Mooneys

Looking to the future, it would seem that Mooneys in general and the new Mooneys in particular are airplanes whose time has come. With the cost of fuel continuing to go ever higher, assuming it is available at all, the high-speed single-engine airplane with an adequate range which burns fuel in miserly amounts will, in many instances, replace light and even medium twins which are not much faster, if faster at all, and which are vastly more expensive to buy, operate and maintain at an only minimal increase in utility. And even the much-touted safety factor of that second engine seems less and less meaningful in the light of the inherent reliability of today's aircraft engines. Also, most non-professional twin-drivers would be hard-put to operate the airplane safely on one engine at any time except during high-altitude cruise. Therefore, it would seem most appropriate that in making the decision to add a new model to its product line, Mooney opted for a 6-place pressurized single-engine aircraft rather than a light or even medium twin.

Table 1-1 shows when the various changes were made in the basic M-20 design and at what price the airplanes were advertised at the time (manufacturer's suggested base price). Also provided is information on the engines used and the speeds claimed in the pilot's handbooks (speed: mph; weight: lbs.).

Chapter 2

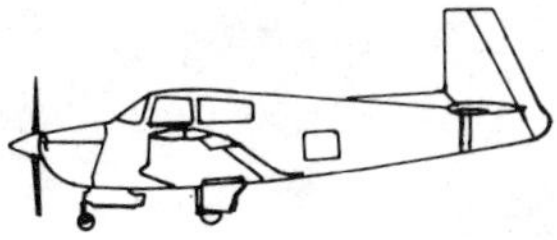

The Mark 21 And The Ranger

The Mooney M-20B, marketed as the Mark 21 (Fig. 2-1) was the first all-metal version of the M-20. It was introduced to the aviation public in 1961. Powered by a 180-hp Lycoming 0-360-A1A engine, it was advertised as having a maximum speed of 190 mph (165 knots) and a cruising speed at 75 percent of power of 182 mph (158 knots). In retrospect, these figures would seem to be somewhat on the optimistic side. The airplane had an empty weight of 1,525 pounds, 110 pounds more than its wood-wing predecessor, with the gross weight having remained unchanged at 2,450 pounds. These figures indicate a useful load of 925 pounds. Based on the 50-gallon usable-fuel capacity (52 gallons total), this results in a payload-with-full-fuel figure of 625 pounds or three 170-pound adults plus 115 pounds of baggage or optional equipment. Of this model, 221 were manufactured before it was replaced the following year by the M-20C—also marketed as the Mark 21. The only difference between the two was an increase in the gross weight by 125 pounds of 2,575 pounds and a minimal change in the engine from the 0-360-A1A to the 0-360-A1D. A total of 1,675 M-20C-s were manufactured and sold under the Mark 21 designation. In 1968, the name of the airplane was changed to the Ranger.

Except for the all-metal construction, this airplane is typically Mooney. Its various systems are designed to be as simple and foolproof as possible. A perfect example of this aim toward simplicity is the famous manually rectractable gear. It is so basic that no backup system is needed. The gear is spring-balanced and hooked up directly to a 2-foot-long lever in the cockpit between the two

Fig. 2-1. The Mooney Mark 21 Ranger.

front seats. When the gear is down, the lever is in a vertical position and locked into a locking device under the instrument panel. When the gear is up, the lever is in a horizontal position on the cabin floor between the seats, and there locked into another locking device. The actual locking mechanism is built right into the handle on the lever. This gives the pilot a firm visual indication of the gear's position and locking. The effort involved in moving the gear either up or down, while certainly greater than that needed with an electrically or hydraulically operated gear, does not require any undue muscle power one you have mastered the trick of knowing which way to grab the handle in order to move the lever in one smooth continuous movement. This design, in addition to making a gear-system failure virtually impossible, has the additional advantage of eliminating the chance of accidentally hitting the wrong switch and, considering the actions necessary to drop the gear, it reduces the likelihood of inadvertent gear-up landings because it is hard to forget or ignore that large handle.

The Flaps

The flap system is another area in which simplicity reigns. The flaps are operated by a hydraulic pump. The handle is below the instrument panel and more or less in the center. To lower the flaps, you push a small control switch down and then, using the handle, you pump several short strokes. Two such strokes result in the right flap setting for takeoff. It takes six or seven strokes to produce full flaps. At that point, the pump handle can no longer be pushed. To raise the flaps, you simply lift the control switch which causes the hydraulics to depressurize while springs pull the flaps to their *up* position. I have always found the system easy to use and thoroughly reliable. It is likely to be more trouble-free than the more sophisticated arrangements used on most other high-performance aircraft.

Rubber Doughnuts

Another unique feature is the use of "rubber-doughnuts" (Fig. 2-2) as shock absorbers in the various gears. Although the rubber does deteriorate with age and must occasionally be replaced, this system eliminates the need to monitor the inflation of oleo struts and the annoying situation which can develop when, by flying from an area with relatively warm weather into subfreezing temperatures, the oleo struts become suddenly underinflated.

Aside from these features and the fact that the fuel gauges are calibrated in pounds instead of gallons (actually a most sensible indication, since pounds, meaning weight, is what we are concerned with when loading an airplane, not gallons, which is, a volume measurement), the airplane has all the other distinctly Mooney features. The wings are a laminar section with tips which are simply cut off flat without the usual curves and the ailerons run all the way to the wingtips. While I have always found this physically attractive, it is said to be at least partially responsible for the heavier-than-normal aileron forces. The tail is an all-moving section. When the trim is moved forward or backward, the entire tail section (horizontal and vertical stabilizers and all) is being moved.

Fig. 2-2. Rubber doughnuts are used as shock absorbers in all Mooneys.

A similar arrangement was later used on the Lockheed JetStar. The advantage, according to aerodynamicists, is that it removes the need for a trim tab and thereby reduces drag. The vertical tail (Fig. 2-3), contrary to the otherwise popular swept-back design, is vertical in front and swept forward in the back. Giving an impression, to me at least, that the airplane is eager to take off and get going.

Leg and Elbow Room

Despite the fact that Mooney has had a reputation (fostered gleefully by its competition) for having an uncomfortably small cabin, the front seats, at least, provide ample leg and elbow room. On the other hand, the back seats do force occupants into some minor contortions in order to find a place for their legs. Long trips with four passengers aboard might not be advisable (also not possible because full fuel cannot legally be carried with four aboard). Getting in and out takes a bit of practice and the right consecutive order. The first persons to enter must be the pilot and the occupant of the right front seat. They then have to pull their seats forward in order to make it possible for the rear-seat passengers to climb in. As is true with most low-wing aircraft, getting into the airplane gracefully is possible only after a considerable amount of practice—if at all. In the beginning, most pilots can usually be observed crawling into the left seat on their hands and feet.

Once inside, you are perfectly comfortable despite the fact that the instrument panel seems unusually close to your face and that there is not much room to move around. The normal seating position is quite similar to that used in a Triumph or other small sports car.

Ground Effect

The airplane sits very close to the ground. It is not particularly difficult to taxi. However, the limited turning radius of the nose wheel makes it difficult to make sharp turns without using differential braking. Takeoff is routine except that the considerable ground effect will cause the airplane to want to lift off before it has achieved actual flying speed. The answer is to either hold it on the ground until the airspeed indicator tells you that climbout can be initiated, or it can be lifted off and flown within inches of the ground while the gear is pulled up and additional speed is gained. This latter procedure is a bit trickier and requires some practice.

The cruise portion of the flight is where the Mark 21 (and, for that matter, all other Mooneys) tends to excel. Unless the airplane is out of rig, a straight-line cross-country cruise is a pleasant experience that requires a minimum of attention to the controls. The noise level is acceptable. There is no noticable vibration. If the flight is made at night, the cockpit lighting, provided by several little spotlights located above and behind the pilot's right ear, is more than adequate. Heating and ventilation are good. However, the latter can get a bit noisy. The entire airplane gives the impression of being a safe and well-constructed piece of machinery. And it is fast, although not as fast as the overly optimistic handbook figures would like to have us believe. Still, with its exceptionally economical 180-hp 4-cylinder engine (one of the most reliable engines ever built by either Lycoming or Continental), it does cover a given distance in less time and on less fuel than any comparable airplane.

Gear Extension

In these early Mooneys, the gear extension speed is an uncomfortably low 120 mph (109 knots). The gear cannot be used as a

Fig. 2-3. The trademark of all Mooneys is the swept-forward tail.

Fig. 2-4. The original window shape of the Mark 20 and Mark 21.

speed brake. Therefore, it is important that the pilot plan the approach early and reduce speed in plenty of time to avoid arriving over the numbers with too much speed. At the right approach speed, the Mark 21 is easy to land. But at too high a speed, it will insist on floating along the runway in ground effect and the only way to get it on the ground before you run out of runway might be to get rid of the flaps. A nose-high landing attitude is exceptionally important and much of the time a slow-flight approach with some power might be better than a power-off glide. The glide ratio of the airplane is 12.7 to 1. That is better than any comparable airplane and is beaten only by gliders.

In 1965, an electrically retracted gear was offered as an option and in 1966 PC (positive control, actually a Brittain wing leveler which was always on but could be overridden by the pilot by depressing a button on the yoke) became standard equipment. In 1968, the Mark 21 was renamed the Mooney Ranger and given a new paint job.

Appearance

While I am on the subject of paint, Mooneys in those days were given some rather startling paint designs and color combinations. Few of which, to my way of thinking, did much to enhance the appearance of those airplanes.

Sometime during that same period, the original shape of the windows (Fig. 2-4) rounded and not particularly graceful, was changed to a more rectangular design. This seems to be more appropriate for the overall shape of the airplane (and it appears to make it look longer and more spacious (Fig. 2-5).

Fig. 2-5. Later the shape of the windows was changed. This made the airplane look longer and more spacious.

Except for the name change, there is really no noticeable difference between that airplane and the Mark 21. At least not until the brief later period when the airplane was produced by Butler. Then the name Mooney was replaced by the name Aerostar and the distinctive tail was changed to a bullet-shaped appendage at the top

Table 2-1. Specifications.

Parameter	Mark 21	Ranger
Engine	L/O-360-A1D	L/O-360-A1D
TBO (hours)	1,200	2,000
Propeller	Hartzell c.s.	Hartzell c.s.
Length	23 ft 2 in	23 ft 2 in
Height	8 ft 4.5 in	8 ft 4.5 in
Wing span	35 ft	35 ft
Airfoil (tip/root)	64-415/63-215	64/415/63-215
Wing area	167 sq ft	167 sq ft
Wing Loading	154. lb/sq ft	15.4 lb/sq ft
Seats	4	4
Empty weight	1,525	1,525
Useful load	1,050	1,050
Gross weight	2,575	2,575
Payload w. full fuel	750	750
Fuel capacity	52 gal	52 gal
Baggage capacity	120 lbs	120 lbs
Takeoff distance	815 ft	1,395 ft
Takeoff over 50-ft obstacle	1,250 ft	1,495 ft
Rate of climb	800 fpm	1,000 fmp
Service ceiling	17,500 ft	17,500 ft
Maximum speed	190 mph	179 mph
Cruise (75% at 7,500 ft)	182 mph	172 mph
Economy cruise (55% @ 10,000)	164 mph	162 mph
Range, max cruise (45 min res)	860 sm	834 sm
Range, econ. cruise(45 min res)	957 sm	930 sm
Stall speedV_{so})	57 mph	56 mph
Landing distance	595 ft	550 ft
Landing over 50-ft obstacle	1,550 ft	1,550 ft

that is vaguely reminiscent of the early Learjets. A total of 395 Rangers was manufactured and sold from 1968 until 1971 when all manufacturing activity was brought to a halt.

After Republic Steel bought the company in 1973, it started production by again building rangers (as well as some of the other previously established models). The name Mooney was returned to the airplane along with the original clean design of the vertical tail. It remained in production until the appearance of the 201 in 1976. Then it was gradually phased out.

See Table 2-1 for a comparison of the Mark 21 and Ranger specifications.

Chapter 3

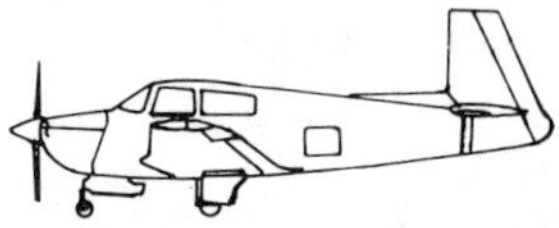

The Super 21 And The Chaparral

In 1964, Mooney started marketing the M-20E under the name Super 21. As far as the airframe, dimensions, cabin and such are concerned, there is no difference between this aircraft and the Mark 21/Ranger. The difference is in the engine. The Super 21, in later years known as the Chaparral, uses a 200-hp fuel-injected Lycoming engine (the IO-360-A1A). However, the 20 extra horsepower does not translate into 20 additional miles in terms of speed.

According to early handbook figures, it is 7 miles faster at maximum speed, and 5 miles faster in cruise. Later handbooks, using somewhat more realistic speed figures, show a difference of 5 miles for maximum as well as cruise speeds.

The gear, flaps and other systems are the same in this airplane as in its 180-hp counterpart. The handling quality is also typically Mooney, on the stiff side, as a result of the fact that solid rods and ball bearing rollers are used in place of the more usual pulleys and cables.

Vapor Lock

Since the engine is very tightly cowled, as is true of all Mooneys, the fuel-injection system is prone to develop a vapor lock when the airplane is shut down at a place where the outside air temperature is high. This situation results when heat soaks the injector system as the airflow through the cowl is reduced to near zero.

One recommended procedure of getting rid of a vapor lock is to turn on the boost pump and flood the engine. This might flush the vapor out of the system. Sometimes this works, sometimes it does not. When all else fails, there is a more cumbersome method which usually does the trick. Soak some rags in cold water, wring them out and then place them on the fuel-injector distributor to reduce its temperature.

Cruise Speed

One thing most Mooney pilots have had to get used to is the fact that the average cruising speed puts the airspeed indicator into the yellow arc. This arc extends from 150 mph to 190 mph. This is well within the range of speeds at which most cruise in the Super 21 is conducted. It's a bit disconcerting at first because the yellow arc means that these speeds are to be used only in smooth air. Every time you hit a bump, you are tempted to slow down. What happens in practice is that most pilots simply ignore that yellow-arc indication altogether. It is not really that crucial, considering that the airplane is built like the proverbial truck. Unless the turbulence is really serious, there is little danger of anything going structurally awry.

Table 3-1. Specifications.

Parameter	Super 21	Chaparral
Engine	L/IO-360-A1A	L/IO-360-A1A
TBO (hours)	1,200	1,600
Propeller	Hartzell c.s.	Hartzell c.s.
Length	23 ft 2 in	23 ft 2 in
Height	8 ft 4.5 in	8 ft 4.5 in
Wing Span	35 ft	35 ft
Airfoil (tip/root)	64-415/63-215	64-415/63-215
Wing area	167 sq ft	167 sq ft
Wing loading	15.4 lb/sq ft	15.4 lb/sq ft
Seats	4	4
Empty weight	1,560	1,600
Useful load	1,015	975
Gross weight	2,575	2,575
Payload w. full fuel	715	675
Fuel capacity	52 gal	52 gal
Baggage capacity	120 lbs	120 lbs
	760 ft	760 ft
Takeoff distance	1,300 ft	1,550 ft
Takeoff over 50-ft obstacle	1,110 fpm	1,400 fpm
Rate of climb	19,500 ft	18,800 ft
Service ceiling	197 mph	190 mph
Maximum speed	187 mph	182 mph
Cruise (75% @ 7,500 ft)	168 mph	158 mph
Economy cruise (55% @ 10,000)	700 sm	772
Range, econ. cruise (45 min res)	882 sm	830
Stall speed (V_{so})	57 mph	57 mph
Landing distance	595 ft	595 ft
Landing over 50-ft obstacle	1,365	1,550

Marketing

The Super 21 has the distinction of having been marketed under a confusing variety of names. It was the Super 21 from 1964 until 1969. In 1969 it became the Chaparral. Then, when Butler bought Mooney and temporarily joined it with the Aerostar, it was sold first as the 201 Chaparral and in 1971 as the Aerostar 201. In 1974, after Reepublic Steel bought what was left of the dormant Mooney company, the Chaparral name was reinstated. The airplane continued to be sold under that name until 1976. Production was then discontinued in favor of the 201.

See Table 3-1 for a comparison of the Super 21 and the Chaparral.

Chapter 4

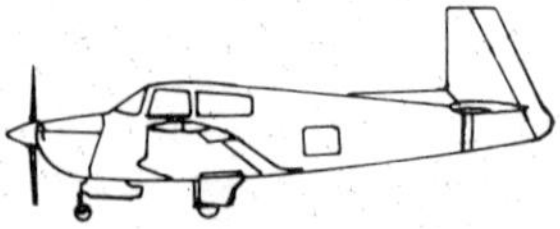

The Executive And The Statesman

In 1967, Mooney was apparently tired of constantly being told that the airplanes were uncomfortably small and began to market a stretched version of the same old M-20. This one was at first the M-20F, whic is known as the Executive, and a year later the M-20G, the Statesman, which proved to be less than successful and production of which was discontinued after two years.

In both instances, the basic Mooney cabin was simply lengthened by about 10 inches to give rear-seat passengers more adequate legroom. In the process, the gross weight, in the case of the Executive, was raised by 165 pounds to 2,740 pounds. The gross weight of the Statesman ended up being actually 50 pounds less than that of the Ranger and Chaparral.

Engines

Again, the difference between the two aircraft is the engine. The Executive uses the same engine as the Super 21/Chaparral—the Lycoming IO-360-A1A which produces 200 hp—while the Statesman is powered by the 180-hp Lycoming 0-360-A1D, also used in the Mark 21/Ranger. Both aircraft are somewhat slower than the comparable smaller versions, but the performance of the Executive is still quite respectable. It is the Statesman which doesn't really come up to snuff. The carbureted 180-hp engine seems insufficient for the larger aircraft. This is especially noticeble when you compare the maximum sea-level rate-of-climb figures for the two aircraft. For the Executive the figures are 1,330 fpm and for the Statesman they are 875 fpm. Since the rate-of-climb

Table 4-1. Specifications.

Parameter	Statesman	Executive
Engine	L/O-360-A1D	L/IO-360-A1A
TBO (hours)	2,000	1,600
Propeller	Hartzell c.s.	Hartzell c.s.
Length	24 ft 3 in	24 ft 3 in
Height	8 ft 4.5 in	8 ft 4.5 in
Wing span	35 ft	35 ft
Airfoil (tip/root)	64-415/63-215	64-415/63-215
Wing area	167 sq ft	167 sq ft
Wing loading	15.1 ob/sq ft	16.4 lb/sq ft
Seats	4	4
Empty weight	1,585	1,622
Useful load	940	1,118
Gross weight	2,525	2,740
Payload w. full fuel	640	746
Fuel capacity	52 gal	64 gal
Baggage capacity	120 lbs	120 lbs
Takeoff distance	815 ft	595 ft
Takeoff over 50-ft obstacle	1,250 ft	1,385
Rate of climb	875 fpm	1,330 fpm
Service ceiling	17,000 ft	17,900 ft
Maximum speed	178 mph	184 mph
Cruise (75% @ 7,500 ft)	169 mph	179 mph
Economy cruise (55% @ 10,000)	?	148 mph
Range, max cruise(45 min res.)	578 sm	591
Range, econ. cruise(45 min res)	?	1,004 sm
Stall speed (V_{SO})	57 mph	62 mph
Landing distance	595 ft	640 ft
Landing over 50-ft obstacle	1,670 ft	1,786 ft

figures in all non-turbocharged aircraft decline drastically at altitude and under high-temperature conditions, this becomes a serious drawback in the case of the Statesman during operations in the mountains and from high-altitude airports.

Thus, while the Statesman limped to early oblivion (only 190 were built), the Executive proved eminently successful. Nearly 1,000 Executives were built and sold before it became the subject of the aerodynamic cleanup which subsequently resulted in the 201 and eventually the 231.

Range

The Executive carries 12 gallons of fuel more than the smaller aircraft. This increases the maximum range at economy cruise to close to 1,000 statute miles. Aside from that difference, the airplane handles just like all the other Mooney airplanes. Even the takeoff and landing distances, at gross weight, are not appreciably different.

See Table 4-1 for a comparison of the Statesman and the Executive.

Chapter 5

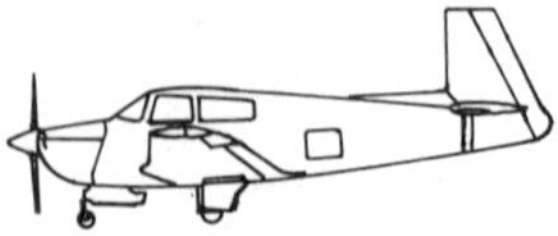

The 201

The 201 and the Turbo 231, although in many ways appearing to be brand new, are the result of a painstaking aerodynamic clean-up job performed on the old Mooney Executive. The basic airframe has remained unchanged, as has the cabin (except for more elegant interior finishing). The wings and tail are the same except that the wing of the 231 has received a new and different tip treatment.

But to say that the 231 is simply a turbocharged version of the 201 would be wrong. The various aerodynamic improvements which managed to milk an additional 20-plus mph out of the normally aspirated airplane, and those which turned the 231 into the exeptional airplane it is, are described in detail in Chapter 7. Here let us look closely at one airplane at a time.

High Performance

The 201 (Fig. 5-1), officially the M-20J, is a 4-seat single-engine, high-perfromance airplane with an electrically operated tricycle landing gear with rubber shock discs, a steerable nosewheel and hydraulically operated disc brakes. It has a 5-foot 11 9/16-inch wheelbase and a turning radius, without differential braking, of 41 feet. The engine is a 4-cylinder, horizontally opposed, air-cooled and fuel injected powerplant with a wet-sump lubricating system that is manufactured by Lycoming. It has a recommended time between overhauls of 1,600 hours and is rated at 200 horsepower at 2,700 rpm at sea level. The displacement is 361 cubic inches and the compression ratio is 8.7 to 1.

The propeller is a constant-speed, hydraulically controlled type with a single-acting governor. It is manufactured by McCauley

Fig. 5-1. The Mooney 201, officially the M-20J, is a cleaned-up version of the Executive.

(a subsidiary of Cessna) and replaces the old Hartzell propeller used on previous Executives. This particular design proved to be more efficient for this application.

The fuel tanks hold a total of 66.5 gallons and 64 gallons are usable. The required fuel grade is 100 or 100LL. Total oil capacity is 8 quarts, with 6 being the minimum permissible for flight.

The gross weight is 2740 pounds and there are no landing-weight or zero-fuel-weight restrictions. The useful load, assuming average equipment, is 1100 pounds. With full fuel, the maximum payload (including the pilot) is 716 pounds. This figures out to four 170-pound persons plus 36 pounds of luggage. See Table 5-1 for the airspeed limitations that apply to the airplane.

The white arc on the airspeed indicator covers a speed range from 61 to 125 mph (53 to 109 knots) and represents to full flap operating range. The lower limit coincides with the stall speed in approach configuration (V_{so}). The upper limit is the maximum permissable speed with the flaps extended. The green arc represents the normal operating range and extends from 68 to 200 mph (59 to 174 knots). The lower limit represents the stall speed in a

Table 5-1. Airspeed Limitations for the Mooney 201.

V_{ne} (never-exceed speed)	225 mph (196 knots)
V_{no} (max structural cruising speed)	200 mph (174 kts)
V_a (maneuvering speed)	135 mph (117 kts)
V_{fe} (max flap-extended speed)	125 mph (109 kts)
V_{le} (max landing gear-extended speed)	150 mph (13) kts)
V_{lo} (max speed for gear extension)	150 mph (13) kts)
V_{lo} (max speed for gear retraction)	120 mph (104 kts)
(max pilot-window-open speed)	150 mph (130 kts)

clean configuration (V_s) and the upper limit is the maximum structural cruising speed. The yellow arc, 200 to 225 mph (174 to 196 knots), represents a speed range in which operations may be conducted only in smooth air. The red line, at 225 mph (196 knots), is the maximum speed for all operations—including dives.

The airplane is stressed for a maximum of 3.8g positive and 1.5 g negative load factor. It is a normal category aircraft approved for VFR and IFR day and night operations (when appropriately equipped), but may not be flown into known icing conditions.

Preflight

Preparatory to flight, the preflight and pre-starting checks are routine. Possible exceptions are that the cowlflaps should be open and the ram-air control should be in the off position.

Engine Starting. *Propeller control*—high rpm; *throttle*—open one quarter; *master switch*—on; *mixture control*—full rich; *electric fuel pump*—on to assure that there is pressure, then off; *mixture control*—idle cut off; *magneto/starter switch*—turn and push to start, release to BOTH when engine starts; *mixture control*—move slowly and smoothly to rich; *oil pressure gauge*—if minimum oil pressure is not indicated within 30 seconds, engine must be stopped and trouble investigated; *throttle*—set for 1,000 to 1,200 rpm.

Flooded Engine. The routine varies somewhat for a flooded engine. *Throttle*—full open; *mixture control*—idle cut off; *electric fuel pump*—off; *magneto/starter switch*—turn to START and push forward; *throttle*—retard when engine starts; *mixture control*—open slowly to full rich; check oil pressure as above.

The Pre-Takeoff and Runup Procedure. This includes all of the usual exercises plus the following. *Annunciator lights*—with the throttle retarded, check the lights by pressing the press-to-test button. *Trim*—there is a slight difference between the correct takeoff setting, depending on the location of the cg. With forward cg, meaning that the back seats are empty, set the trim to the upper portion of the band. With aft cg, meaning with back seats occupied, set it to the lower edge of the band. *Magneto check*—set throttle to 1,900 to 2,000 rpm. Check right magneto, then return to both and allow time for the plugs to clear. The check the left magneto. The rpm drop should not exceed 175 rpm on either magneto and the difference between the two magneto drops should not be greater than 50 rpm. It should be noted that if there is no magneto drop, it might indicate that the magneto is improperly grounded or that there is something wrong with the timing. When in doubt, it is suggested that you repeat the procedure at a leaner mixture and a higher rpm setting. This will either clear up the problem or verify that a deficiency does exist. *Takeoff flaps* are set to 15 degrees.

Takeoff

It is important to remember to always move the throttle slowly and smoothly as the engine is equipped with a counter-weighted crankshaft. Rapid throttle jockeying could detune the counterweights and that might cause overspeeding and possible subsequent engine damage. During the early portion of the takeoff roll, make sure that the engine produces proper full-throttle operation. If there should be a significant indication of rough or sluggish engine response, it would be a reason to discontinue the takeoff. When taking off from a gravel strip, it is important to apply throttle slowly. By doing this, the airplane will start to roll before high rpm is developed and gravel or other loose material will be blown back from the prop area instead of being pulled into the prop (which could result in propeller damage).

For Normal Takeoff. *Electric fuel pump*—on; *throttle*: at 2,700 rpm; *rotation*—lift nosewheel at 71 mph (62 knots) indicated; *climb speed*—82 mph (71 knots); *landing gear*—retract before attaining 120 mph (104 knots) indicated; *flaps*—retract; *electric fuel pump*—off. Check to make sure the engine-driven fuel pump is devlivering adequate fuel pressure. For a maximum-performance takeoff to clear an obstacle, use a climb speed of 76 mph (66 knots) indicated until the obstacle has been cleared.

Climbout

Throttle—26 in hg manifold pressure; *propeller*—2,600 rpm (2,700 rpm for best rate of climb); *mixture*—lean for smooth operation; *cowl flaps*—full open; *airspeed*—105 to 115 mph (91 to 100 knots) (for best rate of climb, use 101 mph (88 knots) indicated at sea level, decreasing to 94 mph (82 knots) indicated at 10,000 feet); *ram air*—open after entering clear air. As in all normally aspirated aircraft, the manifold pressure decreases with altitude and can be restored by gradually opening the throttle.

After climb power has been established and the aircraft is properly trimmed for climb, it is a good idea to once more check that all controls, switches and instruments are set and function properly. A word of warning about the use of ram air: It should always be turned off immediately when indications are that icing conditions might be encountered. Using this unfiltered air when flying in snow or, for that matter, any IFR conditions can be dangerous because snow might accumulate in the fuel-injector impact tubes or moisture can freeze in the inlet passages during icing conditions and result in a drastic loss of power. After leaving icing or other such conditions, do not turn the ram air on again until it is certain that all ice has melted from the aircraft. Also, do not use ram-air in dusty air.

Cruise

When going from climb to cruise, it is advisable to let the airspeed build up to cruise airspeed before trimming the aircraft for level flight and reducing manifold pressure and rpm to the desired cruise setting. Close cowl flaps or, if necessary to keep oil and cylinder-head temperatures within the normal operating range, leave them partially opened by pulling the control aft for approximately 3 inches. When cruising at 75 percent of power or less, always lean the mixture once cruise power has been established. Using the EGT gauge, lean until the indicator shows peak. Then, for best economy, enrich to 25 degrees F on the rich side of peak. For best power, enrich to 100 degrees on the rich side of peak. However, the best power setting will result in a slight increase in speed at the expense of a higher fuel flow and a reduction in range. Any change in altitude requires that the mixture be reset in the same manner as described above. Whenever a change in the power setting is to be made, always follow this simple rule. When increasing power, increase first the rpm, then the manifold pressure. When decreasing power, decrease first the manifold pressure, then the rpm. Always stay within the operating limits set forth in the pilot's handbook.

Do you want to get to your destination as fast as possible? Do you want to be able to travel as far as possible without refueling? Or do you want to pick some reasonable compromise between the two? The cruise and range charts in the pilot's handbook show a wide variety of power settings at different altitudes which produce different results. Using three arbitrary altitudes, 4,000 feet, 8,000 feet and 12,000 feet, the ideal combinations would seem to work out something like this.

At 4000 feet, the highest speed attainable at the least fuel flow for that speed is 186 mph (161 knots). Here you use 2,600 rpm and a manifold presssure setting of 24.4 in hg. This produces 75 percent of power and, when correctly leaned, burns 10.5 gallons (63 pounds) of fuel per hour. The result is an endurance (assuming full tanks) of five hours and 10 minutes, and a range with 45-minute reserve of 956 sm (831 nm).

At that same altitude, if you want to be able to go as far as possible or fly as cheaply as possible, the power setting would be 2000 rpm and 17.2 in hg of manifold pressure which results in 37 percent of power. At this setting you will be moving along at 124 mph (108 knots) true airspeed. You can stay airborne for 10 hours and 14 minutes and the range will be 1,271 sm (1,105 nm). And you will be burning fuel at the rate of 5.7 gallons (34 pounds) per hour.

The compromise between these two extremes would be an rpm setting of 2200 with 22 in hg of manifold pressure. This produces 55 percent of power, resulting in a true airspeed of 161 mph (140 knots), a flight endurance of seven hours and 10 minutes, and a range of 1153 sm (1,003 nm), while burning 7.8 gallons (47 pounds) of fuel per hour.

All figures given above and below assume that the aircraft was at its full 2740-pound gross at takeoff.

At 8000 feet, the numbers change slightly. Here the power setting for the highest true airspeed is 2700 rpm and 23.7 in hg of manifold pressure. The result is 75 percent of power. The speed will be 195 mph (169 knots). You will be able to stay aloft for five hours and the range will be 960 sm (835 nm). The fuel flow will be 10.8 gallons (64.7 pounds) of fuel per hour.

For best endurance, range and economy, the power setting would be 200rpm and 18.2 in hg manifold pressure. The result would be 40 percent of power. The speed would be 135 mph (117 knots), the range 1277 sm (1110 nm), the endurance nine hours and 28 minutes and the fuel flow six gallons (35.7 pounds) per hour.

Here the ideal compromise might be an rpm setting of 2200 , using 20 in hg manifold pressure, resulting in 50 percent of power. The engine would burn 7.3 gallons (44.1 pounds) of fuel per hour and produce a true airspeed of 158 mph (137 knots). You could stay in the air for seven hours and 40 minutes for a range of 1208 sm (1.050 nm).

Now let's look at what happens at 12,000 feet. At that altitude, 75 percent of power is no longer available. The best you can do is 64 percent. This calls for 2700 rpm and 19.8 in hg manifold pressure. This gives you a true airspeed of 185 mph (161 knots). You will be burning 9.6 gallons (57.5 pounds) of fuel per hour. The range will be 999 sm (869 nm), and you can stay aloft for five hours and 40 minutes.

At the other extreme, you can use 2200 rpm and 17.8 in hg of manifold pressure for 42 percent of power. Even at this miserly setting, the true airspeed will be 145 mph (126 knots). You will be burning 6.4 gallons (38.5 pounds) of fuel per hour. You can stay airborne for eight hours and 35 minutes and the range will be 1243 sm (1081 nm).

Somewhere in between these two, the best suggestion might be 2400 rpm and 17.9 in gh manifold pressure. This gives 50 percent of power, a true airspeed fo 165 mph (143 knots), an endurance of seven hours and 20 minutes and a range of 1168 sm (1016 nm). Fuel would be used at a rate of 7.6 gallons (45.5 pounds) per hour.

By putting these figures side-by-side, it would appear that an 8000-foot curising altitude is just about ideal for the 201. Table 5-2 is an example for a 500-sm distance.

As far as the pilot is concerned, flying the 201 is not much different than flying any of the earlier Mooneys—except, of course, that you get there faster and more comfortably. The new seats are wider and seem to hug the body more effectively. While the handbook suggests 15 degrees of flaps for takeoff, this might not be the best idea in strong crosswind situations. The flaps tend to cause the wing to want to fly too early. You might end up skitting across the runway as weight is lifted off the main gear. Assuming sufficient runway—and the 201 is not really an ideal short-field airplane—it is a lot better to forget about the flaps and to take off without them. This makes it easier to keep the wheels firmly on the ground until liftoff speed is reached.

The 201 is a fast airplane and it is not too happy in slow flight. It tends to be a bit skittish and unless you're one of those pilots who engages the autopilot as soon as there is a few inches of air between the wheels and the concrete, you might have to work a bit to keep it straight and level—especially in turbulence. Cruise is the ideal condition for this airplane. The visibility is excellent and the speeds compare favorably with airplanes which are much more expensive. The fuel consumption is such that many an automobile would be hard put to cover a comparable distance at less fuel cost.

The descent and approach are easier in the 201 than in earlier airplanes. This is partly becasue the geardown speed has been raised to 150 mph (130 knots) and partly because the added gear doors, while reducing drag in cruise, increase drag considerably with the gear down. This makes the process of slowing to an acceptable approach speed a lot easier. That doesn't mean that the airplane won't float if it crosses the numbers too fast. It sits as low

Table 5-2. Operating Factors.

	time (hr:min)			fuel (pounds)		
	4000 ft	8000 ft	12,000 ft	4000 ft	8000 ft	12,000 ft
Best speed	2:41	2:34	2:42	169.4	165.9	155.4
Best endurance	4:02	3:42	3:27	137.1	132.2	132.8
Economy cruise	3:06	3:02	3:02	145.9	133.6	137.8

to the ground as did all earlier Mooneys and the ground effect will make itself unpleasantly felt unless the airplane has been slowed to just above stall speed.

Normal Descent and Landing. *Mixture*—rich or lean at pilot's option, as long as smooth operation results; *power*—at pilot's discretion, but the airplane should not be operated for any length of time at between 1500 and 1900 rpm with manifold pressure setting below 15 in hg; *cowl flaps*—closed. Prior to landing: *Landing gear*—extend once the airplane has been slowed to below 155 mph (135 knots); *mixture control*—full rich; *propeller*—high rpm; *fuel*—on fullest tank; *wing flaps*—down 33 degrees when the airplane has been slowed to below 132 mph (115 knots). However, crosswind landings, assuming adequate runway length, might better be made with less or no flaps; *trim*—adjust as needed; *electric fuel boost pump*—on; *ram air*—off.

Go-Around. In the event that a go-around is called for, *Power*—full throttle, 2,700 rpm; *airspeed*—75 mph (65 knots) indicated; *flaps*—retract when climb has been established; *gear*—retract. However, it might be safer to leave the gear down if all that is anticipated is a turn around the pattern prior to another landing attempt. It would be embarrassing, not to mention expensive, to forget to put the gear down again for the second landing attempt; *cowl flaps*—full open.

When landing is assured. *Airspeed on final*—81 mph (71 knots) indicated with full flaps; *touchdown*—on main wheels. Hold the nosewheel off the runway as long as possible; *wing flaps*—retract during taxi; *cowl flaps*—open; *electric fuel boost pump*—off after landing; *trim*—move to takeoff position.

During taxi. *Throttle*—1000 to 1200 rpm. Engine shut down. *Throttle*—idle at 1000 to 1200 rpm until cylinder-head temperature starts to drop; *cowl flaps*—open; *radio master switch*—off; *electrical equipment switches*—off; *mixture control*—idle cutoff; *throttle*—retard as engine stops firing; *magneto starter switch*—off.

One important feature which is new with the 201 and 231 is the annunciator panel. This feature is rarely found on single-engine aircraft. It consists of a variety of warning indicators which light up when a situation develops which calls for pilot attention.

Indicators. The indications are: *gear unsafe*—tells the pilot that the gear is not down or up and safely locked. The pilot should recycle the gear and if that doesn't work, there are instructions in the handbook about the manual emergency operation of the gear; *left or right fuel low*—indicates that between 2.5 and 3 gallons of fuel are left in the respective tank. It is time to switch to the fuller tank (or, if there is no fuller tank, to look for a place to land); *VAC (flashing)*—suction pump is producing less than 3.5 in hg pressure; *VAC (steady)*—suction is above 5.5 in hg, resulting in possibly unreliable readings of the directional gyro and artificial horizon. The system will have to be checked and adjusted or repaired as soon as practical; *volts (flashing)*—low voltage; *volts (steady)*—overvoltage or trippage of voltage relay.

Low voltage is usually caused by an alternator malfunction, belt slippage, or a tripped circuit breaker. If resetting the alternator-field breaker does not succeed in restoring the alternator, all electrical equipment not absolutely essential for flight should be turned off and the fligtht should be terminated as soon as possible. In the event of overvoltage, turn off the radio master switch and then turn the master switch first off, then on to attempt to reset the voltage regulator. If that does not correct the problem, pull out the alternator-Field circuit breaker. Then turn off all electrical equipment not essential to flight and land as soon as possible to effect repairs. *Ram air*—this comes on if the ram-air door is open and the gear is down. Close before landing.

Chapter 6

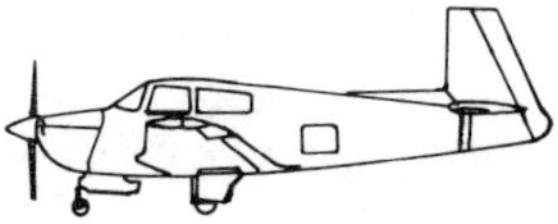

The Turbo 231

Now let's look at the Turbo 231 (Fig. 6-1). Officially it is called the M-20K. The basic description of the 231 is identical to that of the 201—except for the engine. The engine is a 6-cylinder horizontally opposed air-cooled, fuel-injected and turbocharged system. The engine is manufactured by Continental and the fixed-waste-gate turbo charger is supplied by Rajay. It has a recommended time between overhauls of 1800 hours and it is rated at 210 horse power at 2700 rpm at sea level or at any altitude at which sea-level power can be provided by the turbocharger. The displacement is 360 cubic inches and the compression ratio is 7.5 to 1.

The propeller is 2-bladed and roughly the same as used on the 201. It has a slightly different model number and it is manufactured by McCauly. The fuel tanks hold 80 gallons and 72 gallons are usable. The required fuel grade is 100 or 100LL. Total oil capacity is 8 quarts with 5 being the minimum permissable for flight.

The gross weight is 2900 pounds and there are no landing-weight or zero-fuel-weight restrictions. The useful load, assuming average equipment, is 1100 pounds. With full fuel, the maximum payload (including the pilot) is 668 pounds. This equals three 170-pound adults plus 158 pounds of baggage. Table 6-1 lists the airspeed limitations that apply to this ariplane.

The white arc on the airspeed indicator covers speeds from 66 to 125 mph (58 to 109 knots) and represents the full-flap operating range. The lower limit coincides with the full-gross stall speed in approach configuration (V_{so}). The upper limit is the maximum permissible speed with the flaps extended. The green arc repre-

Fig. 6-1. The Mooney Turbo 231, officially the M-20K, is the fastest single-engine airplane on the market today.

sents the normal operating range and extends from 72 to 200 mph (63 to 174 knots). The lower limit represents the stall speed in a clean configuration (V_S), and the upper limit is the maximum structural cruising speed. The yellow arc, 200 to 225 mph (174 to 196 knots), represents the speed range in which operations must be conducted only in smooth air. The red line, 225 mph (196 knots), is the maximum speed for all operations—including dives.

The airplane is stressed for a maximum of 3.8g positive and 1.5g negative load factor. It is a normal category aircraft approved for VFR and IFR day and night operations (when appropriately equipped) but may not be flown into known icing conditions. Sideslips with flaps extended should be avoided at power settings below 25 in hg manifold pressure. At airspeeds above 190 mph (165 knots) indicated, too rapid throttle movement can result in harmful rpm overspeed. Unlike the 201, this airplane has no ram-air capability. The extra boost is provided by the turbocharger.

Flight Preparation

Preparatory to flight, the preflight and pre-starting checks are fairly routine. It should be noted that the fuel tanks are equipped with so-called anti-siphon devices. The fuel-filler neck will trap fuel, giving the impression of a full tank, even when it is nowhere near full. Always push the bottom of the filler open when filling the tanks and when checking the fuel quantity in the tanks. The fresh air vent, unlike vents on other Mooneys, is located in the dorsal fin. It should be checked before flight to assure that no animals or other foreign matter are lodged in it. The cowling has a separate opening where the turbocharger is located. Prior to flight, the turbocharger should be visually inspected to be sure that there are no oil leaks and that all fasteners are secure.

Engine Starting. *Cowl flaps*—open; *throttle*—one-fourth open; *propeller*—high rpm; *mixture control*—full rich; *master switch*—on; *annunciator lights*—test; *primer*—hold on for three to five seconds except at temperatures below 20 degrees F when priming should be continued while cranking until the engine starts. Continuous cranking should never exceed 30 seconds. If the engine has not started by then, several minutes are needed to permit the start to cool before the cranking process can be repeated. *Magneto/starter switch*—turn and push to start, release to BOTH when engine starts. If the engine appears to not want to continue to run, use additional prime; *throttle*—set at 1000 to 1200 rpm; *engine oil pressure*—if minimum oil pressure is not indicated within 30 seconds, the engine should be stopped and the problem investigated; *ammeter*—check.

Flooded Engine. The routine varies a bit for a flooded engine. *Fuel boost pumps*—off; *throttle*—full forward; *mixture*—idle cutoff; *magneto/starter switch*—turn and push. Release to BOTH when engine starts; *throttle*—retard to 1,200 rpm; *mixture control*—full rich; *engine oil pressure*—same as above.

Starting the engine when it is hot calls for the same procedure. However, the throttle is opened initially only very slightly and the mixture is set at full rich from the very beginning.

While taxiing, it might be necessary to increase the rpm somewhat to avoid having the low-voltage light keep on flashing. The pre-takeoff runup procedure is relatively standard except that the engine must not be run at runup speed until the oil temperature has reached a minimum of 75 degrees F. Otherwise, a loss of oil pressure might result. The magneto check is the same as with the 201 with the same limits in terms of individual rpm drop and the relation between the two magnetos.

Table 6-1. Airspeed Limitations for the Turbo 231.

V_{ne} (never exceed speed)	225 mph (196 knots)
V_{no} (max structural cruising speed)	200 mph (174 knots)
V_{a} (maneuvering speed)	135 mph (118 knots)
V_{fe} (max flap-extended speed)	125 mph (109 knots)
V_{le} (max landing gear-extended speed)	150 mph (130 knots)
V_{lo} (max speed for gear extension)	150 mph (130 knots)
V_{lo} (max speed for gear retraction)	120 mph (104 knots)
(max pilot-window-open speed)	150 mph (130 knots)

Takeoff

Always be aware that the turbocharger has no individual oil-temperature indicator. The oil to the turbocharger is provided by the engine oil and its temperature will be the same. In order to assure proper turbocharger operation and to prevent possible damage, the oil must be at least 100 degrees F before takeoff. Since adequate lubrication is of paramount importance in turbocharger operation, the oil-temperature indication must be watched carefully, especially in cold weather, as the engine must under no circumstances be operated at high power with an oil temperature indication below 100 degrees F.

Because of the action of the turbocharger, the manifold pressure will increase to 40 in hg when the throttle is approximately two-thirds open. The throttle must not be advanced farther since harmful overboost might occur. If manifold pressure, rpm and fuel flow are in their respective maximum-power ranges, maximum power is being developed even though the throttle is not fully open. During the winter, in cold weather and at low field elevations, 40 in hg manifold pressure might be reached with even less throttle. Always monitor the manifold pressure gauge when applying takeoff power. The turbocharger is of the always-on type with a fixed wastegate. It will provide extra boost under all altitude and density altitude conditions. That is why the overboost potential exists. Though such overboosting should be carefully avoided, a momentary overboost above 40 in hg, but not exceeding 43 in hg for a period of under 10 seconds, will not harm the engine.

Takeoff Procedures. *Fuel boost pumps*—off; *alternate air*—closed; *engine-oil temperature*—100 degrees F or higher ; *power*—40 in hg manifold pressure and 2,700 rpm; *rotation*—lift the nose wheel at 74 mph (64 knots); *landing gear*—retract before reaching 120 mph (104 knots); *flaps*—retract in climb. A manifold pressure of 40 in hg can be maintained during the climb, but must not be exceeded. Do not lean the mixture during climb unless power is reduced for one reason or another to below 75 percent. The compressor-discharge-temperature gauge (CDT) should be monitored. If CDT limits appear to be exceeded, a power reduction is indicated.

Normal Climb. *Power*—33 in hg manifold pressure and 2600 rpm; *mixture control*—full rich; *cowl flaps*—open as required to keep temperatures within allowable limits; *airspeed*—110 to 130 mph indicated (96 to 113 knots).

Best Rate Of Climb. *Power*—40 in hg manifold pressure and 2700 rpm; *mixture control*—full rich; *cowl flaps*—open; *airspeed*—110 mph (96 knots) at sea level, decreasing to 95 mph (83 knots) indicated above 15,000 feet.

Cruise. Manifold pressure and rpm settings can be determined by consulting the appropriate tables in the pilot's handbook. Cowl flaps might have to be kept partially open to maintain oil temperatures within the normal operating range and cylinder-head temperatures between 250 and 440 degrees F. Continuous operation at a turbine-intake temperature (TIT) in excess of 1650 degrees F is prohibited under all circumstances except that leaning the mixture to a TIT of 1725 degrees F for a maximum of 30 seconds is permissible in order to determine peak TIT. For all practical purposes, turbine-intake temperature (TIT) is the same as exhaust-gas temperature (EGT). The mixture is leaned in the same manner as described for the 201 except that TIT is monitored instead of EGT.

The Oxygen System

One parameter has not been mentioned so far. The current models of the Turbo 231 have a built-in oxygen system (the early models did not). It goes without saying that if high-altitude operations are anticipated or if the flight is a long one during which high altitudes might have to be reached in order to negotiate a weather system, even though high-altitude flight had not initially been planned, the oxygen system should be full and should be checked before flight is intiated. It is always a good idea to have the oxygen system recharged after any amount of use. Oxygen is not available at every airport and oxygen exhaustion would rob the airplane of some of its greatest advantages.

Flight Altitudes

With the 231, as with any turbocharged aircraft, the choices of flight altitudes and of the speeds and fuel flows which can be achieved at the various altitudes are increased considerably. Winds usually blow at much higher velocities at altitudes above 15,000 feet and their effect on the flight profile will (or, at least, should) play a serious role in deciding at which altitude to fly. Inveterate smokers, of course, have no such choice. There can be no smoking in the cabin when oxygen is being used.

In the following paragraphs, the performance parameters for several different altitudes and wind conditions are examined. The altitudes and wind velocities chosen are 4000 feet with a 5-knot head- and tailwind, 8000 feet with a 10-knot head- and tailwind 12,000 feet with a 15-knot head-and tailwind, 16,000 feet with a 20-knot head- and tailwind, 20,000 feet with a 30-knot head- and tailwind, and 24,000 feet with a 40-knot head- and tailwind.

4000 Feet

At 4000 feet, the best speed which can be attained is 160 knots (184 mph) at 75 percent of power. This calls for 2300 rpm and 34.6 in hg manifold pressure. The fuel flow at this power setting is 10.9 gallons (65.4 pounds) of fuel per hour. The endurance is five hours and 48 minutes and the range (with 45-minute reserve) is 924 nm (1063 sm). With the 5-knot headwind, the ground speed is 155 knots (178 mph). An equally strong tailwind brings the ground speed up to 165 knots (190 mph). Assuming a trip profile, the cruise portion of which is 500 nm (575 sm), the distance will be covered in three hours and 13 minutes with a headwind. There will be 35.1 gallons (210.6 pounds) of fuel used, while with a tailwind the time aloft is three hours and two minutes and the fuel consumption is 33 gallons (198.2 pounds) total.

Under the same altitude and wind conditions, the best endurance speed is 120 knots (138 mph). This equals 40 percent of power with 2000 rpm and 24.2 in hg manifold pressure. Fuel is burned at 6.1 gallons (36.6 pounds) per hour. The endurance is 11 hours and five minutes with a maximum range (with reserve) of 1,160 nm (1,334 sm). The headwind ground speed is 115 knots (132 mph) and the tailwind groundspeed is 125 knots (144 mph). The 500-nm distance will therefore be covered in four hours and 21 minutes on 26.5 gallons (159 pounds) of fuel, or in four hours on 24.4 gallons (146.4 pounds) of fuel, depending on the direction of the wind.

The best compromise at this altitude would seem to be 55 percent of power which results in 140 knots true airspeed (161 mph) at 2000 rpm and 29.7 in hg manifold pressure. Here the fuel flow is 8.1 gallons (48.6 pounds) per hour, the endurance is seven hours and 45 minutes and the range is 1075 nm (1236 sm). The head- and tailwind speeds are 135 knots (155 mph) or 145 knots (167 mph) respectively. The time to cover the 500-nm distance is reduced to three hours and 42 minutes on 30 gallons (180 pounds) of fuel or three hours and 27 minutes on 28 gallons (168 pounds) of fuel.

8000 Feet

At 8000 feet, this time with 10-knot wind components, the highest speed is 167 knots (192 mph) which requires 75 percent of power, 2350 rpm and 33.8 in hg manifold pressure. The fuel flow is up to 11 gph (66 pph), the endurance is five hours and 35 minutes and the range is 940 nm (1081 sm). Headwind reduces the speed to 157 knots (181 mph), while tailwind increases it to 177 knots (204 mph). The 500-nm trip will take three hours and 11 minutes against the wind or two hours and 49 minutes with the wind. Fuel consumption is either 35 gallons (210 pounds) or 31 gallons (186 pounds).

At this altitude, the best endurance speed is 123 knots (141 mph) at 2000 rpm and 23.4 in hg manifold pressure. The result is 40 percent of power. The fuel flow is 6.1 gph (36.6 pph), the endurance is 10 hours and 55 minutes and the range comes to 1164 nm (1339 sm). Ground speed, depending on the direction of the wind, is either 113 knots (130 mph) or 133 knots (153 mph). This translates for the 500-nm trip into either four hours and 25 minutes on 27 gallons (162 pounds) of fuel or three hours and 46 minutes on 22.9 gallons (137. 4 pounds) of fuel.

The 55-percent compromise brings the true airspeed up to 145 knots (167 mph) at 2100 rpm and 28.2 in hg manifold pressure while using 8.2 gph (49.2 pph). The endurance is down to seven hours and 38 minutes and the range is 1100 nm (1265 sm). The ground speed, adjusted for the head- and tailwind components, will be either 135 knots (155 mph) or 155 knots (178 mph) and the trip will take either three hours and 42 minutes or three hours and 13 minutes while using 30.3 gallons (181.8 pounds) or 26.4 gallons (158.4 pounds) of fuel respectively.

12,000 Feet

Now we climb up to 12,000 feet where the wind blows at 15 knots. Here the best speed is 172 knots (198 mph) at 75 percent with 2,450 rpm and 32 in hg manifold pressure. Fuel is burned at 11.2 gph (67.2 pph). Th endurance is five hours and 30 minutes and the range is again 940 nm (1,081 sm). The high and low ground speeds are 157 knots or 187 knots (181 or 215 mph). This means that the 500-nm trip will take three hours and 11 minutes against the wind, burning 35.6 gallons (213.6 pounds) of fuel, while with the wind it will take only two hours and 40 minutes and use 29.9 gallons (179.4 pounds) of fuel.

The best endurance figures look like this. At 40 percent of power, using 2000 rpm and 22.5 in hg manifold pressure, the true airspeed is 124 knots (143 mph) while using 6.1 gph (36.6 pph). You can stay in the air for 10 hours and 43 minutes and cover a distance of 1165 nm (1340 sm). The ground speed, adjusted for the wind components, is either 109 knots (125 mph) or 139 knots (160 mph). The trip will take four hours and 35 minutes on 28 gallons (168 pounds) of fuel against the wind or three hours and 36 minutes on 22 gallons (132 pounds) of fuel with the wind.

At this altitude, 55 percent of power, using 2250 rpm and 25.9 in hg manifold pressure, produces 150 knots (173 mph) TAS. With the prevailing wind, this translates to ground speeds of 135 knots (155 mph) or 165 knots (190 mph). The endurance is seven hours and 30 minutes and the range is 1125 nm (1294 sm). The time and fuel figures for the 500-nm trip are three hours and 42 minutes or three hours and two minutes on 30.7 gallons or 25.1 gallons (184.2 or 150.6 pounds) of fuel.

16,000 Feet

At 16,000 feet, the wind has increased to 20 knots. Here the best speed, at 75 percent, is 178 knots (205 mph) at 2550 rpm and 30.6 in hg manifold pressure. Fuel is burned at 11.3 gph (67.8 pph). The endurance is five hours and 27 minutes and the range is 950 nm (1093 sm). The ground speed is either reduced to 158 knots (182 mph) or increased to 198 knots (228 mph) depending on your direction of flight relative to the wind. The trip times come to three hours and 10 minutes or two hours and 32 minutes. Fuel consumption would be either 35.7 or 28.6 gallons (214.2 or 171.6 pounds).

Looking at the best endurance figures, the speed is 125 knots (144 mph) at 2200 rpm and 20 in hg manifold pressure. This equals 40 percent of power. Fuel is burned at 6.3 gph (37.8 pph). Maximum time in the air is eight hours and 40 minutes, and maximum distance is 1124 nm (1293 sm). Ground speeds will be 105 knots (121 mph) against the wind and 145 knots (167 mph) with the wind. Resulting trip times are four hours and 46 minutes or three hours and 27 minutes. Therefore, the fuel used will be either 29.9 or 21.7 gallons (179.4 or 130.2 pounds).

At the more usual 55 percent, the speed is 155 knots (178 mph) while 2350 rpm and 24.5 in hg manifold pressure. Fuel flow is 8.5 gph (51 pph), endurance is seven hours and 14 minutes and the range is 1075 nm (1236 sm). With or against the wind, the ground speeds are either 135 or 175 knots (155 or 201 mph). Resulting trip times are three hours and 42 minutes or two hours and 38 minutes. Fuel cost is 31.5 or 24.3 gallons (189 or 145.8 pounds).

Now we get to those altitudes where the turbocharged airplane really comes into its own. But here the winds are usually considerably stronger. At an altitude of 20,000 feet, the wind component is assumed to be 30 knots. Here the best true airspeed is 186 knots at 75 percent, using 2650 rpm and 29.8 in hg manifold pressure, resulting in a fuel-flow figure of 11.2 gph (67.2 pph), an endurance of five hours and 18 minutes and a range of 960 nm (1104 sm). The ground speed, against the wind, is down to 156 knots (179 mph). With the wind it is up to 216 knots (248 mph). It will take three hours and 13 minutes to fly 500 nm against the wind and two hours and 19 minutes to cover the distance with a tailwind. Fuel used will be either 36 or 25.9 gallons (216 or 155.4 pounds).

The figures for best endurance speed are 45 percent of power, 2400 rpm, 20.4 in hg manifold pressure and 135 knots (155 mph) TAS. Fuel is burned at 7.3 gph (43.8 pph); endurance is seven hours and 40 minutes and range is 1070 nm (1231 sm). Adjusted for the wind components, the ground speeds are either 105 or 165 knots (121 or 190 mph). Resulting trip times are either four hours and 46 minutes or three hours and two minutes. Fuel use is either 34.7 or 22.1 gallons (208.2 or 132.6 pounds).

At 55 percent, the TAS is 160 knots (184 mph), using 2500 rpm and 23.4 in hg manifold pressure, burning 8 gallons (48 pounds) of fuel per hour. Endurance is six hours and 40 minutes for a range of 1040 nm (1196 sm). Head- and tailwinds change the speed to a ground speed of 130 or 190 knots (150 or 219 mph), resulting in trip times and fuel consumption of three hours and 51 minutes on 30.8 gallons (184.8 pounds) of fuel or two hours and 38 minutes on 21 gallons (126 pounds) of fuel.

24,000 Feet

And now to the highest altitude for which the aircraft is certificated. It can go much higher, but the certification is limited to that altitude because, in the event of a malfunction of the oxygen system, the pilot flying at a much higher altitude would only have moments of useful consciousness during which to dive down to an altitude at which oxygen starvation is no longer a dangerous problem. Here, at 24,000 feet, the wind component is assumed to be a conservative 40 knots. The best speed at this altitude is 192 knots (221 mph) while using 2700 rpm and 28 in hg manifold pressure for 74 percent of power. Fuel flow is 11.1 gph (66.6 pph), endurance is five hours and 10 minutes and range is 950 nm (1093 sm). The ground speed, with or against the wind, is 152 or 232 knots (175 or 267 mph). Resulting trip times are three hours and 17 minutes on 36.5 gallons (219 pounds) or two hours and 10 minutes on 24 gallons (144 pounds).

Best endurance speed at this altitude calls for 45 percent while using 2550 rpm and 19.8 in hg manifold pressure for a TAS of 140 knots (161 mph). Fuel flow is 7.5 gph (45 pph), the endurance is seven hours and 15 minutes and the range is 1025 nm (1179 sm). Ground speed against the wind is 100 knots (115 mph) and with the wind is 180 knots (207 mph). Against the wind, the trip will take five hours and use 37.5 gallons (225 pounds) of fuel. With the wind, the trip will only take two hours and 47 minutes at a fuel cost of 20.9 gallons (125.4 pounds).

At 55 percent at that altitude, the TAS is 174 knots (200 mph) while using 2650 rpm and 22.4 in hg manifold pressure. Fuel is burned at 8.8 gph (52.8 pph). The endurance is six hours and 30 minutes and the range is 1005 nm (1156 sm). With 40-knot winds, the ground speed will be either 134 or 214 knots (154 or 246 mph). The 500-nm distance, in the event of a headwind, will be covered in three hours and 44 minutes on 32.8 gallons (196.8 pounds). In the event of a tailwind, it will only take two hours and 20 minutes and use 20.6 gallons (123.6 pounds) of fuel.

In order to put all this multitude of figures into perspective, see Table 6-2. This chart shows that, in terms of fuel consumed, the cheapest way to fly the 500-nm trip is to climb to 24,000 feet and to use 55 percent of power—assuming a 40-knot tailwind, On the other hand, if you are fighting the various headwind components at the different altitudes, you actually get there on the least fuel at 4000 feet while using the best endurance speed of 40 percent. The best ground speed against the wind is achieved at 16,000 feet, using 75 percent of power, while at 55 percent the ground speed remains the same at 4000, 8000, 12,000 and 16,000 feet. With the wind in your back, the best ground speed (as was to be expected) is at 24,000 feet while using 74 percent of power. At 55 percent, it is also at the same altitude.

The primary purpose of all of this is to dramatically demonstrate the importance of picking the right altitude and the appropriate power setting to either minimize the detrimental effect of headwinds or to take full advantage of the help offered by tailwinds.

Handling Characteristics

The handling characteristics of the Turbo 231 are not dramatically different from those of other Mooneys. However, the ailerons seem to be lighter and more effective in terms of roll rate. Supposedly, this is the result of the new and different wing-tip treat-

Table 6-2. Turbo 231 Performance Parameters.

	knots	rpm	in hg	% power	gph	endurance	range	kts/tailwind	kts/headwind	500 nm time/headwind	500 nm time/tailwind	500 fuel/headwind	500 nm fuel/tailwind
4,000 ft													
Best speed	160	2300	34.6	75	10.9	5:48	924	155	165	3:13	3:02	35.1	33.0
Best endurance	120	2000	34.2	40	6.1	11:05	1160	115	125	4:21	4:00	26.5	24.4
Economy cruise	140	2000	29.7	55	8.1	7:45	1075	135	145	3:42	3:27	30.0	28.0
8,000 ft													
Best speed	167	2350	33.8	75	11.0	5:35	940	157	177	3:11	2:49	35.0	31.0
Best endurance	123	2000	23.4	40	6.1	10:55	1164	113	133	4:25	3:46	27.0	22.9
Economy cruise	145	2100	28.2	55	8.2	7:38	1100	135	155	3:42	3:13	30.3	26.4
12,000 ft													
Best speed	172	2450	32.0	75	11.2	5:30	940	157	187	3:11	2:40	35.6	29.9
Best endurance	124	2000	22.5	40	6.1	10:43	1165	109	139	4:35	3:36	28.0	22.0
Economy cruise	150	2250	25.9	55	8.3	7:30	1125	135	165	3:42	3:02	30.7	25.1
16,000 ft													
Best speed	178	2550	30.6	75	11.3	5:27	950	158	198	3:10	2:32	35.7	28.6
Best endurance	125	2200	20.0	40	6.3	8:40	1124	105	145	4:46	3:27	29.9	21.7
Economy cruise	155	2350	24.5	55	8.5	7:14	1075	135	175	3:42	2:38	31.5	24.3
20,000 ft													
Best speed	186	2650	29.8	75	11.2	5:18	960	156	216	3:13	2:19	36.0	25.9
Best endurance	135	2400	20.4	45	7.3	7:50	1070	105	165	4:46	3:02	34.7	22.1
Economy cruise	160	2500	23.4	55	8.0	6:40	1040	130	190	3:51	2:38	30.8	21.0
24,000 ft													
Best speed	192	2700	28.0	74	11.1	5:10	950	152	232	3:17	2:10	36.5	24.0
Best endurance	140	2550	19.8	45	7.5	7:15	1025	100	180	5:00	2:47	37.5	20.9
Economy cruise	174	2650	22.4	55	8.8	6:30	1005	134	214	3:44	2:20	32.8	20.6

ment which is also credited with a reduction in adverse yaw. Rudder pressures are very light. This is a good thing, too, considering that there is no rudder trim. Stalls, regardless of the configuration, are honest stalls but without much buffet. There seems to be no tendency to drop out of even accelerated stalls and roll-control is satisfactory right down to the actual stall. All in all, the 231 is somewhat easier to handle in slow flight than any of its predecessors.

One thing must be kept in mind by pilots who like to fly way up there at 24,000 feet or so. There is always the possibility that because of oxygen exhaustion, or for some other reason, a fast descent must be executed. There are two fairly satisfactory means of accomplishing this. One is to dive the airplane clean and achieve speeds right up to the red line of 225 mph (196 knots). But this should not be tried in rough air. As a matter of fact, since it is impossible to know whether the air is smooth at the different altitudes, it's probably a bad idea anyway. However, it does result in a rate of descent of somewhere around 2000 fpm. By far, the better alternative is to slow the airplane to 150 mph (130 knots) and put the gear down. This will keep speed of the plane in a range which is safe even if turbulence is encountered on the way down. This, too, can produce a rate of descent of 2000 frpm.

During normal descents, the most important consideration is to keep the engine from cooling too rapidly since it might then refuse to accelerate satisfactorily when power is applied. The trick is to use sufficient manifold pressure, preferably above 23 or so inches, which will tend to keep the cylinder-head temperature in the green. On descents from high altitudes at something like 1000 fpm, it is obviously necessary to plan the descent early enough to avoid arriving over the destination with thousands of feet of excess altitude. A normal descent from 24,000 feet to a sea level airport will take nearly a half an hour and use up some 90 to 100 miles in the process!

Descent. *Mixture*—as required for smooth operation; *throttle*—as required; *propeller*—for long descents use minimum governing rpm and a manifold pressure no lower than is needed to maintain the desired performance; *cowl flaps*—closed; *cylinder-head temperature*—no lower than 250 degrees F.

During the approach, it will be noticed that the 231 is less squirrely in gusty conditions at low speeds than the 201. Therefore, it is somewhat easier to land.

Before Landing. *Mixture*—full rich; *propeller*—high rpm; *fuel boost pump*—off; *wing flaps*—at pilot's discretion, but full down at speeds below 125 mph (109 knots); *trim*—trimming should be accomplished throughout the approach as speed is being reduced. The force required for nose-up position when flaps are lowered increases considerably. Adequate trimming will reduce this problem. Lowering the gear has little effect on the trim problem. Lowering the gear has little effect on the trim position. Make sure that the *parking brake* is completely off.

Go-Around. In the event of a go-around *power*—40 in hg manifold pressure and 2700 rpm; *wing flaps*—move to takeoff position as soon as climb is established; *trim*—trim nose down; *airspeed*—accelerate to 91 mph (77 knots) indicated; *landing gear*—retract unless the go around is to be followed immediately by another landing. Gear and gear doors produce considerable drag; *cowl flaps*—open.

Landing. *Airspeed on final*—86 mph (75 knots) indicated with full flaps. For short-field landing use an approach speed of 80 mph (69 knots) indicated with full flaps. For landings in a crosswind exceeding 12 knots, use half flaps or, in extreme cases, no flaps at all.

Table 6-3. Specifications.

Parameter	201	231
Engine	L/IO-360-A3B6D	C/TSIO-360-GB-1
TBO (hours)	1,600	1,800
Propeller	McCauley c.s.	McCauley c.s.
Length	24 ft 8 in	25 ft 5 in
Height	8 ft 4 in	8 ft 4 in
Wing span	35 ft	36 ft 1 in
Wing area	167 sq ft	175 sq ft
Wing loading	16.4 lb/sq ft	16.6 lb/sq ft
Seats	4	4
Empty weight	1,640	1,800
Useful load	1,100	1,100
Gross weight	2,740	2,900
Payload w. full fuel	716	668
Fuel capacity (usable)	64 gal.	72 gal.
Baggage capacity	120 lbs	120 lbs
Takeoff distance	900 ft	1,200 ft
Takeoff over 50-ft obstacle	1,560 ft	2,100 ft
Rate of climb	1,030 fpm	1,080 fpm
Service ceiling	18,800 ft	24,000 ft
Maximum speed	225 mph	225 mph
Cruise (75% @ 7,500 ft)	195 mph	190 mph
Cruise (75% @ 24,000 ft)	NP	221 mph
Economy cruise (55% @ 10,000)	170 mph	175 mph
Economy cruise (55% @ 24,000)	NP	187 mph
Range, max cruise (45 min res)	789 sm	775 sm
Range, econ. cruise (45 min res)	957 sm	1,294 (max @ 55%)
Stall speed	66 mph	68 mph
Landing distance	870 ft	1,135 ft
Landing over 50-ft obstacle	1,975 ft	2,270 ft

Taxi. *Trottle*—100 to 1200 rpm; *cowl flaps*—full open; *trim*—takeoff position.

Accessories

For both the 201 and 231, Mooney offers an intimidatingly wide choice of factory-installed accessories and equipment. This includes avionics from King, Bendix or Collins, complete integrated flight-director/autopilot systems, area navigation, radar, radar altimeter, radio-telephone, RMI, IVSI, and fuel management computers, and, for the 231, there is a propeller-deicing option. Judging by the large number of purchasers who are willing to pay the extra price in terms of dollars as well as payload to acquire all of these goodies, the theory among many seems to be that as long as the airplane has roughly the same capability as a twin at two or three times the price, the money saved in purchase cost and operational economy might just as well be spent on the best that the avionics and equipment manufacturers have to offer.

See Table 6-3 for a comparison of the 201 and the 231.

Chapter 7

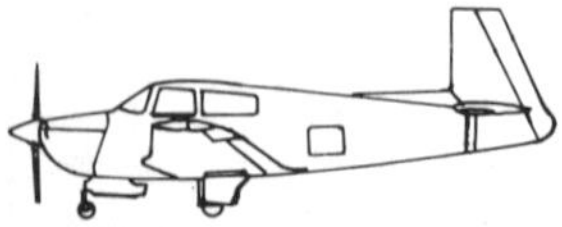

Making An Airplane Go Faster

Normally, when a company decides to make an airplane that goes faster, the immediate reaction is to use a bigger engine. A perfect example of this philosophy is the Piper Comanche which, during its life span, went from 180 to 250, then to 260 and 400 hp. The trouble with this approach is that the cost of fuel burned increases quite out of proportion to the amount of speed gained. But, as has since been proved, there are other ways to achieve the goal of higher speeds without increasing the fuel flow at the same time.

Roy Lopresti

To find out how this minor miracle was wrought, the logical person to talk to is Roy Lopresti (Fig. 7-1), vice president for research and development of Mooney Aircraft Corp.

"I guess I better start at the beginning. Republic bought the company on October 3, 1973. And at that point, they started to look for people to run the company and they got me from Grumman. I arrived here on the last day of 1973, December 31. And just between me and the wall, if I had known then how difficult this was going to be, I might not have come, because, by comparison to what I had been used to at Grumman, there wasn't anything here, absolutely nothing. I have a very warm feeling in the pit of my stomach when I see the airplanes coming off the line now. I feel very proud of the fact that now we are a profit-making company, and that I played some part in making it happen, because during the first two years our backlog each month was like 15 airplanes, and we had barely enough orders for the next month. But 15 airplanes is not a

Fig. 7-1. Roy Lopresti, vice-president for research and development, Mooney Aircraft Corp.

break-even number, so we were losing money each month (Fig. 7-2). During that first year, in 1974, we built Chaparrals, some Rangers, some Executives, and ended each month in the red, but we had, of course, known that this was going to happen anyway. And the second year, 1975, was not that much better.

"But, aside from being responsible for getting the airplanes into production, my personal job was to think downstream toward the next product. And from the very beginning, as soon as we had an engineering department, one engineer, we started thinking about things that would make the airplane go faster. The fuel shortage had started the year before and we were all kind of keen on making the airplane more fuel efficient. And, of course, everyone likes to make an airplane go faster. But to make it go faster and at the same time make it more efficient—that was the way to point your product. In other words, what we were looking for was to make the airplane go faster without putting in more horsepower. They nicknamed me 'Speed Freak' because I always wanted everything to go faster. The whole point is that if you go faster but don't burn any more fuel, then you are more efficient. So, from the very beginning we started doing things to make the airplane go faster. I was only one of two pilots on board at the time, so I did most of the flying. We did things like we built a set of inboard gear-doors. They were kind of jury-rigged; we had cables in the cockpit to pull them up tight. And within a month's time we knew what they were going to be worth in terms of speed. We kept doing those things and just kept them in the back of our heads. We played with wing tips, we played with wing-root fillets, and we were also doing things with engines.

"During all this time we had to keep the company running. So the R and D was kind of a part-time job. And then 1975 came along and we had to decide how we were going to spend money for the following year. Don Cox (director of marketing) of course, wanted a bigger airplane, a faster airplane, a better airplane, everything that marketing always wants. So I was looking at higher horsepower—a Tiara engine. We flew a Mooney Executive with 285-hp Tiara, and with a 320-hp Tiara and we flew it with a 220-hp Franklin, and we were designing an Executive with a 260-hp engine. And while we were doing that and while, at the same time, I was playing with draft reduction items, I proposed to Don Cox that he could have either one, an airplane with a new engine, or an airplane with cleaned-up drag. And he said, 'Well, how much speed can you give me?' I said, 'How much speed do you want?' He answered, I remember it very clearly, 'Give me 11 miles per hour. That's enough for me. That's almost as much for me, marketing-wise, as having a slightly bigger engine.'

200 mph On 200 hp

"I knew, of course, that a change in speed accomplished by reducing drag was much less expensive. There is always a risk when you change a piece of the airplane, some structure or an aerodynamic piece, there's nearly always some problem that you didn't foresee and that can get you into trouble. But change the drag of the airplane, and the chance of trouble developing is very small. So, anyway, he said he'd be satisfied with 11 mph and as we were proceeding in 1975 we got 11 miles very quickly. And one of the fellows who works with me, Dick Jarvis, he came with me from Grumman and runs the experimental shop. 'Well', Dick said, 'Why don't we do 200 mph on 200 hp?'

"That set us to thinking. They would come out in 1976—when the country was 200 years old. It sounded good; 200th birthday—200 mph—200 hp. We'd bring it out on the 4th of July. Great! So we went ahead and announced that plan to the company, even though we didn't really know if we could make it happen.

"The speed before had been 180 mph. Now, based on records we checked, all the people before us tried many times to make the airplane go faster. They had all kinds of plans and programs, and made all kinds of attempts, but they never found a mile an hour.

Fig. 7-2. The backlog each month was about 15 airplanes—not enough to break even (courtesy of Mooney).

And I think I know why. As simple as it may sound, knowing how fast an airplane goes is really kind of tricky. What happens in most small companies is that you plot a curve for altitude and speed. You fly your airplane at wide-open throttle and you get a bunch of different speeds which may give you a nice smooth curve on your chart. Then you bring the airplane back to the hangar and make a change, a very small change. Then you fly it the next day and the atmosphere has changed a little bit and you get a different curve, and you say, 'I don't know whether it's faster or slower. I really can't tell.' The reason is that the horsepower changes because of changes in the temperature, and with a small change you simply may not be able to discern the difference properly. So we had to work out a system where we could eliminate the effects of the atmosphere.

"We hired a leading aerodynamicist, or rather a team, husband and wife, Fenn and Dorothy Taylor, and they put together a test procedure that is not really new to the industry as a whole, but it's new in the small-plane business. What we do is, instead of flying a series of altitudes, we fly one altitude but at a series of speeds and chart the data. These data get reduced right back to lift and drag coefficients which eliminates all the temperature variables. Meanwhile, we make the change and then go back and fly the airplane again. You can compare the lift and drag coefficients for the two cases very carefully. We can actually discern differences in speed on the order of one mile per hour very accurately. And if you

Fig. 7-3. The air inlet on the older Mooney was a big gaping hole.

make enough changes worth one mile per hour and add them up, you get a whole big change. That was the basis on which we were able to find a lot of little pieces which, when all stacked together, made a big difference.

"We announced kind of unashamedly that we would make the airplane go 200 mph on 200 hp. Several magazines printed that, and lots of people sort of smiled or sneered. As a matter of fact, several respected experts, one who had been with NASA, said, 'Your can't do that. It's impossible to do.' I had four people working with me on this, and the more people said that we couldn't do that, the more determined we became. We simply decided that we would make it happen, period.

Modifications

"Here are some of the things that we did. The biggest single change, the one that caused the biggest increase in speed, was essentially cooling drag."

The inlet (Fig. 7-3) on the Executive is just a huge gaping hole. And there is a drag-producing installation for the oil cooler, plus a recessed landing light (Fig. 7-4). It was obvious from the very beginning that here was a place where lots of drag reduction could be achieved. The new front (Fig. 7-5) of the airplane is smoother. The air intake is 70 percent smaller while providing the same amount of cooling. The landing light is covered with a smooth lens (Fig. 7-6) and a smaller and very clean ram-air intake was designed. The combination of these changes were worth a whop-

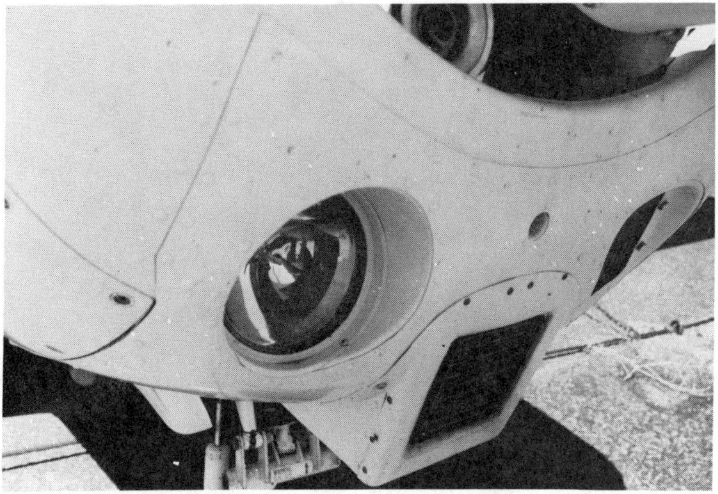

Fig. 7-4. The recessed landing light caused a lot of drag.

ping six miles an hour. It was, in fact, not a very complicated improvement and, in addition to the speed increase, it made the airplane look more modern.

Another early modification was the landing gear. In the Executive, the entire wheel assembly, brake lines and brake systems are exposed to the airflow (Fig. 7-7) and part of the tire is outside the wing contour. All this was obviously creating a fair amount of drag. To solve this problem, an inboard gear door was developed which worked better than expected. Before deciding on the new design, other gear doors, such as the ones on the Bonanza,

Fig. 7-5. The new front of the 201 is smooth and the air intake is 70 percent smaller without any loss of cooling capacity.

Fig. 7-6. The new landing light is covered with a smooth lens.

were examined. But with them, the actuator failed to drive the system and, rather than change all the linkages and mechanisms, it was decided that the new design was an easy alternative which was nearly as good as an entirely separate door. There is still a small portion of the tire exposed, but its effect appears to be minimal, judging by the fact that the modification was good for four miles. And it proved to have another advantage. As Lopresti points out: "Often times, when you do something to get some desired change,

Fig. 7-7. On the old Mooneys, the gear doors left all of the tire and most of the gear system exposed to the air flow.

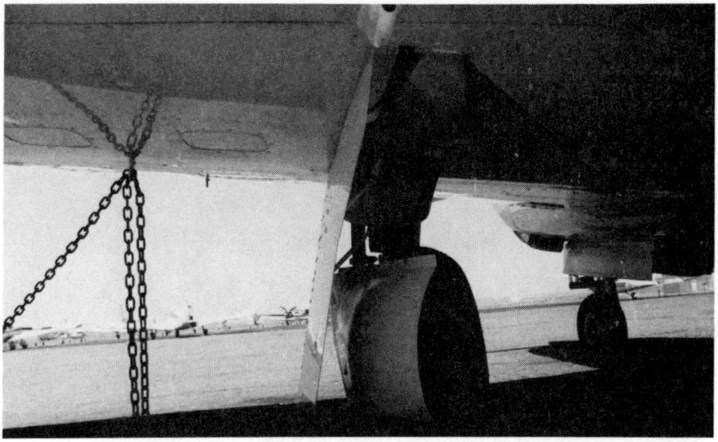

Fig. 7-8. The new gear doors reduce drag in cruise, adding 4 mph to the cruise speed, but increase drag in the approach configuration when added drag is a desirable feature.

you get something else you didn't want. But sometimes it's good. In this case, when the gear is down, the gear door creates additional drag (Fig. 7-8). And the airplane needs drag in the landing configuration. So now the airplane can fly a steeper approach."

In addition, by beefing up a few pieces in the gear system they were able to raise the gear-down speed to 150 mph. That is a very desirable improvement in so fast and clean an airplane.

The exhaust pipe (Fig. 7-9) on all older Mooneys hangs down at a very large and visually unattractive angle. By replacing it with

Fig. 7-9. The exhause pipe on older Mooneys hangs down at a large and unattractive angle.

Fig. 7-10. The new exhaust pipe is elegantly swept and gains one mph for the airplane.

one that is elegantly swept back so that it rides in the shadow of the cowl flap, looks were improved and one mile an hour was gained (Fig. 7-10).

About the famous Mooney wing (Fig. 7-11), Lopresti comments: "It's built fantastically well, but it's built for inexpensive production. It takes twice as many man hours to build the Executive than it does Piper to build an Arrow." But the wing has a profusion of access panels and, while on older Mooneys they were quite smooth and flush riveted, there was apparently a decision made around 1967 to save some money. Lumpy panels and rivets were substituted, apparently in the belief that drag on the underside, of the wing is of no appreciable consequence. On the 201, all

Fig. 7-11. The famous Mooney wing is fantastically strong but expensive to manufacture (courtesy of Mooney).

Fig. 7.12. The old OAT gauge stuck out of the windshield. It looked funny and created considerable noise.

of these are again replaced by flush panels and rivets, making for a perfectly smooth wing.

Lopresti and his group tufted the whole airplane. Wherever they could see tufts flopping around, they knew that something was disturbing the airflow and that improvements were called for. One example was in the area of the horizontal stabilizer where there were openings between it and the tailcone. Some wipers were installed to seal off those gaps and another two miles per hour were gained.

One item which has little if any effect on the speed, but was definitely in need of improvement, was the OAT gauge (Fig. 7-12).

Fig. 7-13. A new sensor has made it possible to place the OAT gauge onto the instrument panel. This left the windshield clean and eliminated the noise factor.

Fig. 7-14. Older Mooneys used to have a fixed step.

An inexpensive $19 item, it stuck out of the windshield, looked funny and created noise. It was decided that the right place for an OAT gauge is on the instrument panel. A sensor was installed and the gauge repositioned despite the fact that the new system cost about $100 (Fig. 7-13).

Years ago, Mooneys used to have a retractable step that cost about $300. That was later replaced by a fixed step (Fig. 7-14). In the 201, this step has been eliminated to save a couple of pounds and up the speed another 1.5 mph. The step is now an option, and though a few 201s are sold with it, most owners prefer that extra mile and a half to the minimal increase in convenience.

Even the tail skid came in for treatment (Fig. 7-15). The old two-piece assembly, neither attractive nor aerodynamically clean, was replaced with a simple one-piece assembly that serves as skid as well as tiedown ring (Fig. 7-16).

The decision that a certain model of propeller, manufactured by McCauley (a subsidiary of Cessna) would be more efficient than the Hartzell prop on the Executive, resulted in a fairly expensive modification. The IO-360 engine used in the Executive was found to be unable to efficiently interact with the new prop, and a different version of the IO-360 had to be substituted. It cost around $200 extra and increased the speed by only very little, but it was considered a worthwhile improvement.

The Last Few Changes

When all the more obvious changes had been accomplished, Lopresti and his crew would simply look at the airplane and search

Fig. 7-15. Even the tail skid, a two-piece affair, came in for the treatment.

for places where there might be unnecessary drag (Fig. 7-17). He admits that it might not have been very scientific, but it paid off in the long run. They made fairings that covered the flap hinges (Fig. 7-18, and reduced the gaps (Fig. 7-19) between the wings and control surfaces—especially at the flaps. There was concern that this might adversely affect the stall speed. But, in practice, the stall speed was found to have actually gone down a bit. The combination off all those little, apparently meaningless, improvements netted another three miles per hour.

Fig. 7-16. The new one-piece tail skid serves the same purpose and is better looking. It also reduces drag by a minute amount.

Fig. 7-17. The engineers would look for gaps and other items that might produce drag.

So, early in 1976, they had succeeded in getting the 201 cleaned up to a point where it would do 195-mph. They were working extra hard to find those last few places that would give them the magic 200-mph figure. Then, in May of that year, with the editor of *Flying* magazine about to come out to fly the airplane, they put on their last fix and in the ensuing test flight got 202.5 mph. Hurray!

Having found an additional 20-plus miles in an airplane that had been pretty fast to start with, it was only logical that the thoughts of this small group of miracle workers would turn to turbocharging. Turbocharging is efficient because at any given percent of power the engine will use the same amount of fuel, turbocharged or not, except that with turbocharging you can obtain that given percent of power at higher altitudes—where the true airpseed is better.

"But," explains Lopresti, "there is no free lunch, I guess. If I take the 201 and turbocharge it and make no other changes, then all

Fig. 7-18. Fairings were designed and manufactured to cover the flap hinges.

I gain is an overall average of something like 20 miles an hour, because I'd be increasing the weight of the airplane by the weight of the turbocharger, additional fuel, an oxygen system and so on. The

Fig. 7-19. The gaps between the wing and the control surfaces were reduced.

Fig. 7-20. The cowling of the Turbo 231 is longer and still more streamlined than that of the 201.

result would be that below the critical altitudes the non-turbocharged airplane would actually be faster."

Still Faster

With this in mind, it was decided to take the 201 and make it still faster by reducing drag even further. In addition, a change in engine was indicated. The Lycoming in the 201 produces 200 hp plus or minus three percent. The 6-cylinder Continental, which was selected for the new airplane, produces 210 hps minus zero or plus 5 percent. As a result, the Continental-powered airplane is just about as fast at sea level as the 201.

Aside from the engine change, the improvements designed to make the airplane go faster include a new, longer and still more streamlined cowling (Fig. 7-20). The wing tips (Fig. 7-21), which has been a Mooney trademark from the beginning, actually proved to produce considerable drag. So a new, upswept wing tip was designed. During tuft testing, the tufts lay straight back. The new tip (Fig. 7-22) resulted in better low-speed control, a speed increase of 1½ mph and greatly superior aileron effectiveness.

Fig. 7-21. The squared-off wing tips had been a trademark of all Mooneys from the very beginning.

Fig. 7-22. The new streamlined wing tips, containing all navigation lights, brought an improvement in low-speed control, better roll response and a speed increase of 1.5 mph.

"And then we started looking again for more little things because we were running out of big things to do. We managed to close the remaining gaps (Fig. 7-23) between the wing and the flaps and ailerons. At the wing-leading-edge, and intersection (Fig. 7-24), we designed a nicely curved, very streamlined cover which proved to be worth another one and a half miles. But that also produced a negative reaction on the stall characteristic. The inner wing, which is supposed to stall first, wouldn't. So we played with some stall strips and managed to solve the problem. Actually, as the result of the two strips (Fig. 7-25) the stall behavior of the 231

Fig. 7-23. All remaining gaps were effectively closed.

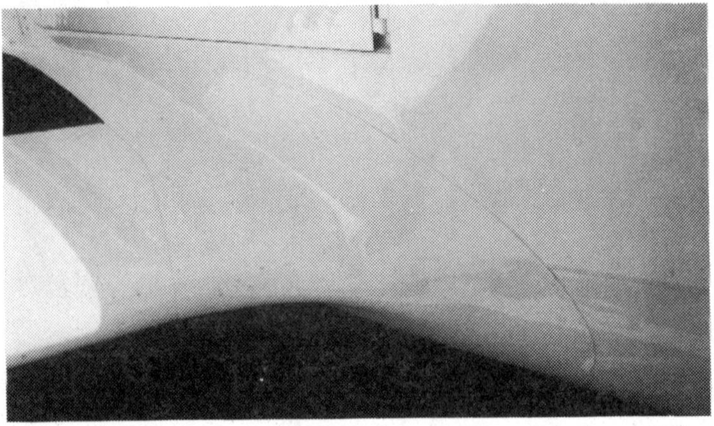

Fig. 7-24. The intersection of the wing leading edge and the fuselage underwent effective streamlining treatment. This accounted for another increase in speed.

is much better than that of the 201. The 201 will occasionally try to roll off on you. The 231 does not."

Like the earlier Mooneys, the 201 had a cabin-air inlet at the top of the fuselage (Fig. 7-26) which, when open, produced drag and was quite noisy. In the 231 (Fig. 7-27), a NASA scoop was placed into the dorsal fin where it results in virtually no drag and is much more quiet. Also, with the old dorsal fin, when the tail is trimmed nose down, there is a big gap between it and the leading edge of the vertical stabilizer. A type of skirt was designed which

Fig. 7-25. The two stall-strips make the stall characteristics of the 231 superior to those of the 201.

Fig. 7.26. On older Mooneys and on the 201 the cabin-air intake is on top of the fuselage, causing drag when it is open, and making a lot of noise.

Fig. 7-27. On the 231, the cabin air-intake is a NASA scoop in the dorsal fin.

Fig. 7-28. The 201 is a cleaned-up version of the Executive.

Fig. 7-29. The Turbo 231—not just a turbocharged 201—but a radically improved airplane. It is the best in its class.

was attached to the dorsal fin and which effectively closes that gap regardless of the position of the trim. And other little improvements include another NASA scoop at the bottom to feed cooling air to the battery, and something as simple as covering the elevator-hinge-bolt holes resulted in a speed increase of two miles.

All in all, it would seem that one might be justified in saying that the transformation of the Executive into the 201 (Fig. 7-28) and 231 (Fig. 7-29) represents a unique chapter in the annals of aviation. Luckily, for all concerned, it took place just at a time when the pressing need for fuel efficiency has greatly increased the acceptance of single-engine airplanes as a means of efficient and economical air transportation.

Chapter 8

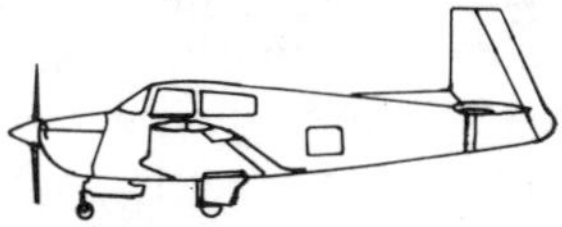

A Comparison Of The 201 And The Turbo 231

There are many ways to compare airplanes. Initial purchase price, speed, comfort, range, service, ceiling, fuel consumption—all these and many other parameters tend to be of varying importance to individual aircraft owners. The initial purchase price depends primarily on the amount of money which is available or which can be borrowed from a bank or other lending institution. Since airplanes tend to depreciate very little, they can often be sold after five, eight or even 10 years for a price which comes reasonably close to the amount paid for the airplane when it was new. Granted, with high inflation, this amount will be worth less then than it is now, but that doesn't change the fact that the number of dollars involved in the cost of ownership, excluding the fixed and variable expenses of operation, maintenance, insurance and so on, is actually quite low.

Comfort. Personal comfort falls into two categories. One concerns the pilot, when flying alone, and the other concerns the passengers. Comfort is usually judged primarily by the amount of room, especially leg room, available for each person. This varies considerably with the physical size of each of the persons involved, but it must be remembered that the larger the cabin, the more money it costs in terms of horsepower and, in turn, fuel to move this cabin through the air at a given speed.

Speed. This is a favorite means of comparing airplanes and the ability to fly fast is one of the primary sales points for almost any airplane. The importance of speed points for almost any airplane. The importance of speed is highly overrated. A three- or five-knot difference in speed adds or subtracts only a few minutes to or from the average mission. This is usually at a penalty in fuel flow which is quite out of proportion for the amount of time saved.

Range. This is important only to those who fly long distances with any frequency. If virtually all flights are over distances of 500 nm or less, none of the more popular airplanes today would have to put down along the way for a fuel stop. However, with fuel availability becoming increasingly uncertain, it is often advantageous to be able to fly to a destination and then return without having to refuel before getting back to home base.

Fuel Flow. In the old days, fuel flow was usually considered to be of minor importance. That has changed drastically. Today, with avgas costing $2 per gallon or more (and jet fuel being not much cheaper) the amount of fuel burned in order to go a given distance is becoming one of the major considerations in determining the overall efficiency of a given aircraft.

Service Ceiling. This is the altitude at which an aircraft at full gross can still climb at 100 fpm and it is something that is rarely taken into serious consideration. For one thing, virtually any airplane, expecially when below gross, is capable of flying considerably higher than its official service ceiling. For another, flying at those altitudes requires the use of auxiliary oxygen, not to mention the fact that above 18,000 feet all operation must be IFR, and all aircraft must comply with minimum equipment requirements.

Where ceiling figures do become important is primarily in the case of twins. Here it is not the service ceiling but the single-engine ceiling. In other words, the altitude which can be maintained in the event that one engine is inoperative. This figure is unacceptably low for most normally aspirated twins, which is why we have only included turbocharged twins in the comparison chart (Table 8-1).

Comparable Aircraft

In order to get a reasonably clear idea of how the Mooney 201 and 231 stack up against other more or less comparable aircraft, we have prepared a chart that lists 17 different aircraft side-by-side. These 17 aircraft fall into four groups. The top six are normally aspirated single-engine retractables. The next seven are turbocharged single-engine retractables. And then, even though they are not actually in the same class except possibly in terms of speed, there are three turbocharged twins and one pressurized single.

They are as follows—201: Mooney; Arrow IV: Piper; Skylane RG: Cessna; Sierra: Beech; Aries: Bellanca; Lance II: Piper; Turbo Arrow IV: Piper 231: Mooney; Turbo Skylane RG: Cessna; Turbo Viking: Bellanca; Turbo Centurion: Cessna; Turbo LanceII: Piper; Bonanza 36TC: beech; Seneca II: Piper; Turbo Skymaster: Cessna; Aerostar 601B: Piper; P. Centurion: Cessna. See Figs. 8-1 through 8-7.

Table 8-1 has 12 columns. The first gives the model designation of the aircraft being compared. The second column headed: *Estimated cost, fully equipped* gives the approximate dollar figures in 1980 dollars which would be involved in purchasing the aircraft with all equipment necessary for efficient operation, both VFR and IFR. The next column, headed: *Seats* shows the number of seats in the airplane, including the pilot's seat. The fourth column, *Payload with full fuel*, gives the weight, in pounds, which can be legally loaded into the airplane when all tanks are full. The fifth column, *Range (nm), economy cruise*, gives the range in nautical miles with allowance for taxi, takeoff, climb and 45 minutes reserve—assuming the aircraft is flown at an economy-cruise setting of approximately 55 percent of power. Column number six, *Speed, knots*, shows the book speed for each of the aircraft at that power setting. And column number seven, *Fuel flow (pph)*, shows the fuel flow in terms of pounds per hour, based on information provided by the various manufacturers.

Now it gets a bit more complicated. The eighth column, *Time 250, 500, 750, 1000*, shows the time in hours and minutes which it will take to cover 250, 500, 750 and 1,000 nautical miles at that same economy-cruise speed. No allowances are made for taxi, takeoff, climb etc., but a 45-minute fuel stop is included for aircraft that have an insufficient range to cover either 750 or 100 nautical miles without refueling.

Fig. 8-1. The Mooney 201 (courtesy of Mooney).

Table 8-1. Performance and Economy Comparisons.

MODEL	extimated cost, fully equipped	seats	payload w. full fuel	range(nm) economy cruise	speed knots	fuel flow (pph)	time 250 500 750 1000	fuel (lbs) 250 500 750 1000	fuel @$2 per gal 250 500 750 1000	cost per nm in $s	cost per seat mile, seats full ($s)
201	70,000	4	716	929	152	52	1:38 3:18 4:56 7:19	84.76 171.6 256.36 341.64	28.25 57.20 85.45 113.88	.1144	.0286
Arrow IV	72,000	4	725	875	131	48	1:55 3:48 5:43 8:29	92.16 182.4 274.56 371.04	30.72 60.80 91.52 123.68	.1216	.0304
Skylane RG	75,000	4	801	1135	139	59	1:47 3:36 5:24 7:57	105.02 202.4 318.6 424.8	35.01 67.47 106.20 141.60	.1349	.0337
Sierra	79,000	4	720	686	115	48	2:10 4:18 7:16 9:27	104.16 206.4 312.96 417.6	34.72 68.80 104.32 139.20	.1376	.0344
Aries	83,000	4	699	917	165	62	1:31 3:01 4:32 6:46	94.24 196.54 280.86 382.54	31.41 65.51 93.62 127.51	.1310	.0328
Lance II	95,000	6	1068	714	128	71	1:47 3:36 6:09 7:57	126.38 255.6 383.4 564.45	42.13 85.20 127.80 188.15	.1704	.0284
Turbo Arrow IV	76,000	4	674	825	139	55	1:48 3:35 5:24 7:56	99.0 196.9 297.0 394.35	33.00 65.63 99.00 131.45	.1313	.0328
231	78,000	4	668	1135	163	50	1:32 3:06 4:36 6:08	76.5 155.0 230.0 306.5	25.50 51.67 76.67 102.17	.1033	.0258

Turbo Skylane RG	80,000	4	661	915	140	64	1:47 3:34 5:22 7:53	113.92 228.48 343.68 456.32	37.97 76.16 114.56 152.11	.1523	.0381
Turbo Viking	97,000	4	371	723	147	69	1:42 3:24 5:51 7:33	117.3 234.6 351.9 469.2	39.10 78.20 117.30 152.11	.1564	.0391
Turbo Centurion	100,000	6	1118	874	152	74	1:38 3:18 4:56 7:19	120.62 244.2 364.82 486.18	40.21 81.40 121.61 162.06	.1628	.0271
Turbo Lance II	105,000	6	781	720	132	83	1:53 3:47 6:26 8:20	156.04 313.74 471.44 629.14	52.01 104.58 157.15 209.71	.2029	.0349
Bonanza 36TC	135,000	5	723	675	157	75	1:35 3:11 5:32 7:07	118.5 238.5 358.5 477.75	39.50 79.50 119.50 159.26	.2390	.0478
Seneca II	150,000	6	743	849	157	108	1:35 3:11 4:47 7:07	170.64 343.44 516.24 687.96	56.88 114.48 172.08 229.32	.2290	.0382
Turbo Skymaster	152,000	6	580	989	157	117	1:35 3:11 4:47 7:07	184.86 372.06 559.26 745.29	61.62 124.02 186.42 248.43	.2480	.0413
Aerostar 601B	245,000	6	838	1129	198	144	1:09 2:17 3:26 4:35	165.6 328.32 493.92 659.52	55.20 109.44 164.64 219.84	.2189	.0635
P. Centurion	165,000	6	864	841	155	76	1:37 3:14 4:50 7:12	123.12 245.48 367.08 490.2	41.04 81.83 122.36 163.40	.2447	.0408

Fig. 8-2. The Piper Arrow IV (courtesy of Piper).

The ninth column, *fuel (lbs) 250, 500, 750, 1000*, shows the amount of fuel being burned, figured in pounds, by each of the aircraft in the process of covering the four different distances. If there appear to be slight inconsistencies in these figures, they are the result of rounding off fractions below 1/100 of a pound. The figures do not make allowances for taxi, takeoff, climb and so on, and are based on the economy-cruise fuel-flow figures supplied by the manufacturers in their owner's manuals.

The 10th column, *Fuel @ $2 per gallon*, shows the cost of this fuel for each of the four distances. The 11th column shows the cost per natucial mile—extrapolated from the 500-nm flight. And the last column divides that figure by the number of available seats to arrive at the cost figure per person on board when all seats are full. It must be remembered that this figure might be deceiving in some instances. Not all of the aircraft can operate with full fuel when all seats are occupied. Those which are fuel-limited might require several fuel stops over the longer distances. This would not change the cost figure for the actual cruise portions of the overall flight, but would add more fuel consumption because of the repeated need to land, taxi, take off, climb, and so on.

In each group we have shaded the boxes which contain the best figures in each category. Thus, we can see at a glance which airplane exceeds in which parameter.

Fig. 8-3. THe Cessna Skylane RG (courtesy of Cessna).

Fig. 8-4. The Beech Sierra (courtesy of Beech).

Let's take column 2: Price. The Mooney is the least expensive in that group. The number of seats (column 3) is the same for all except the Lance II, which can seat two more than the others and, as a result, also has the highest payload with full fuel (column 4). Figuring that each person with luggage somes to about 200 pounds, we find that the only one of the listed aircraft which can legally fill all seats and all tanks at the same time is the Skyland RG. All others would have to leave some fuel out if all seats are to be occupied.

Fig. 8-5. The Piper Lance II and Turbo Lance II (courtesy of Piper).

The Skylane also does best in the range category (column 5). It is the only one which can fly over 1,000 nm without refueling. In the speed category) (column 6) Bellanca's Aries beats the competition. According to the manufacturer, it is 13 knots faster than the next fastest—the Mooney 201. In the fuel-flow category (column 7), the Arrow IV and the Sierra tie for the lowest consumption per hour. The Mooney 201 is the nest lowest. And, assuming that Bellanca's speed figures are correct it takes the Aries less time to cover all four distances than it does any of the others. The Mooney 201 is right on its heels.

Cost and Distance

But now comes the real test. What's it going to cost to fly those distances? Here the Mooney 201 is way ahead of everybody else. It uses less fuel for each trip length, less fuel per nautical mile and costs the operator $13.63 less over the 1,000-nm distance than the Aries at an expense of 33 additional minutes in the air. Actually, the second most economical aircraft is the Arrow IV, but the difference in time aloft, over a 1,000 nm, is one hour and 10 minutes. The only dollar figure in which the Mooney 201 does not excel is in the cost-per-seat-mile column because the Lance can carry six people. But even here, the difference is one 2/100 of a cent. All in all, then, it would seem that the 201 is by a considerable margin the most economic and most efficient aircraft in the group. The necessary compromises to achieve that economy are minor indeed.

Turbocharged Singles

Now let's look at the turbocharged singls. In this group, the Turbo Arrow IV would cost roughly $2,000 less than the Mooney Turbo 231, comparably equipped, and three of the listed aircraft can seat more people. Two of them have six seats and one has five.

Fig. 8-6. The Cessna Turbo Centurion (courtesy of Cessna).

Fig. 8-7. The Beech Bonanza 36TC (courtesy of Beech).

Also, the Turbo Centurion has the highest payload-with-full-fuel figure in the group, but none can actually carry a full fuel load when all seats are occupied. But that is where the competition ends. From here on, in all the remaining parameters, the Mooney Turbo 231 is far ahead of the pack. It has the longest range, the highest speed, the lowest fuel flow, uses the least amount of time and, in turn costs less money to fly all of the listed distances. It is cheaper on the basis of cost per nautical mile and cost per seat-mile than any of the non-turbocharged singles (and, of course, any of the other airplanes listed on the chart). With Mooney's new book figures having been generally acclaimed as being among the most realistic and reliable in the industry, the computations made here would seem to be not overly optimistic.

Turbocharged Twins

In the third group, the turbocharged twins, the Seneca is the least expensive and burns the least fuel. The vastly more expensive Aerostar excells in all other parameters. However, it must be remembered that for the twin-pilot who is less than proficient in engine-out operation, the Turbo-Skymaster is a much simpler to fly and a much safer machine than any other twin on the market today (not counting the Rutan Defiant, which is not yet in production). The twins, and the last airplane, the Pressurized Centurion, are listed here only in order to provide a means of comparing the cost of operating two engines or the only currently available pressurized single to that of operating the vastly more economical, but not much slower, singles.

Overall, the fastest single is supposedly the Aries. The Mooney 231 is two knots slower. The fastest twin, by a country mile, is the Aerostar. The longest range figures belong to the Skylane RG and the Turbo 231. The Aerostar is the only other one with a 1,000-plus-nm range. But dollar-for-dollar, the 231 beats the others all up and down the line.

Chapter 9

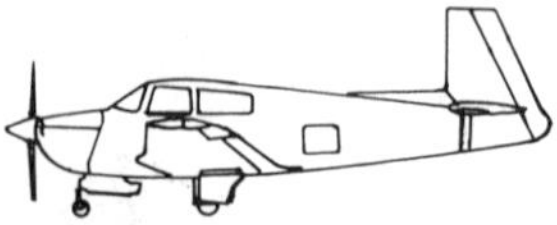

A Comparison Flight

It's one thing to compare different airplanes by simply using the book figures provided by the manufacturers, it's quite another to actually fly different airplanes under comparable weather conditions over the same route. In order to find out how each airplane might handle comparable conditions along a typical flight involving mountainous country as well as the plains, we have laid out a route from Wichita, Kansas, to Grand Junction, Colorado. This is a straight-line distance of 545 nautical miles or an along-the-airways distance of 552 nautical miles.

Since it would be impractical to fly all 17 airplanes used in the comparison chart (Table 8-1) we've simply selected those which seem to be most competitive in terms of dollar-for-dollar operation. They are, of course, the 201 and Turbo 231, plus the two Piper Arrows, the Arrow IV and the Turbo Arrow IV.

Payload

Each airplane has a sufficient payload capability to carry the pilot and two passengers for the entire distance—assuming relatively reasonable wind conditions. And, for the purposes of this comparison, we will assume that all four airplanes started at the same time (9:30 a.m. CST) from Wichita's Mid-Continent Airport, each carrying comparable navcom equipment and each, therefore, following more or less the same route (Fig. 9-1) which coincides with Victor-73, Victor-10, Victor 244 and Victor 484. As stated above, following these Victor airways adds only seven nm to the

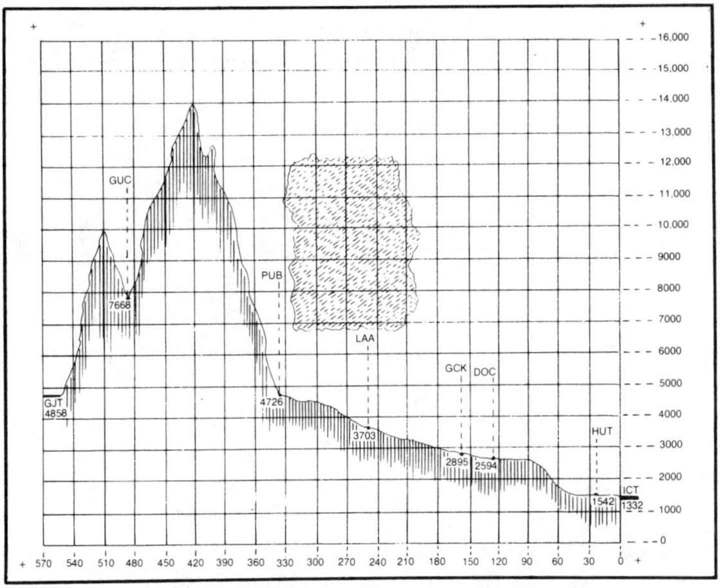

Fig. 9-1. A compressed profile of the topography and the weather along the route from Wichita to Grand Junction.

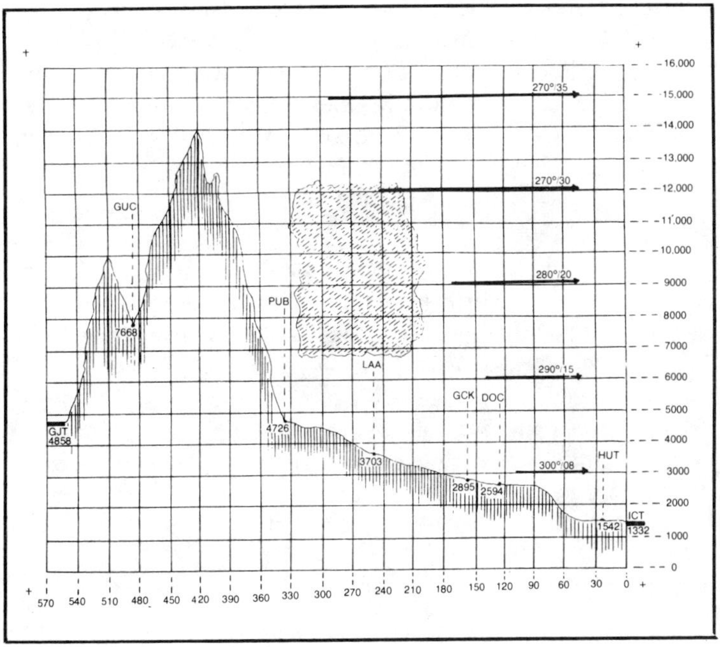

Fig. 9-2. The upper winds as forecast for the trip.

straight-line distance, and it would therefore seem silly to try to avoid the very minor doglegs involved.

There are, of course, two ways of flying this trip. One would be IFR and the other VFR. The forecast weather conditions call for VFR conditions all along the route. There are CAVU conditions between Wichita and Lamar, 3,000-foot overcast between Lamar, and some 50 nm east of Pueblo, with tops at 12,000 feet. Then CAVU again all the way to Grand Junction except for a few scattered thunderstorms in the afternoon—which you might expect to easily avoid. The forecast upper winds along the route are given as 300°/08 at 3,000 feet, 290°/15 at 6,000 feet, 280°/20 at 9,000 feet, 270°/30 at 12,000 feet and 270°/35 at 15,000 feet. See Fig. 9-2.

IFR

If the flight is to be made IFR, all four aircraft have very little choice in the matter of altitudes, except for the stretch between Lamar and Pueblo where a decision must be made whether it is preferable to fly in the clouds for a distance somewhat in excess of 100 nm, or whether the pilot woudl prefer to climb to VFR conditions on top. The MEAs along the route are 3,400 feet msl between Wichita (ICT) and Hutchinson (HUT). From there it rises to 4,300 feet to Dodge City (DOC) and to 4,600 feet to Garden City (GCK). After Garden City, the MEA is 5,600 feet to Lamar (LAA) and then 7,000 feet to Pueblo (PUB). That is the last nav aid before climbin over the ridge of the eastern Rockies. The MEA from Pueblo to Gunnison (GUC) is a healthy 16,000. For the last leg to Grand Junction (GJT) it drops to 14,000 feet. See Fig. 9-3.

Service Ceilings

The service ceilings of the four aircraft are ample enough to permit them to climb to that 16,000-foot MEA. Here are the limits: the Arrow IV: 17,000 feet; the Mooney 201: 18,800 feet; the Turbo Arrow IV: 20,000 feet; and the Mooney Turbo 231: 24,000 feet. Of course, oxygen will be needed, and while the two turbocharged aircraft are equipped with an oxygen supply which is ample for the relatively short time during which it will be needed by all aboard, we will assume that the two non-turbocharged models are carrying sufficient auxiliary oxygen to do the trick.

VFR

If the flight is to be made VFR, there are two basic alternatives. Either we can stay low until we get to Pueblo, and then start to climb to an altitude which will permit us to fly through the mountain passes at an altitude sufficiently low to eliminate the need for oxygen, or we can climb straight to 12,500 feet and then fly at that altitude until it's time to descend toward Grand Junction.

The 201

So, let's start with the 201. Looking at the weather reports and the winds-aloft forecasts, we would probably be tempted to decide to stay as low as possible for the longest period of time and only get up high into those strong headwinds when it comes time to cross the mountains. The four most likely means of flying this trip are shown in Fig. 9-4. The .... line represents the VFR flight at low altitude and the ——....——....——line represents VFR at 12,500 feet. The ------- line represents IFR at altitudes which include a climb into and then cruise in the clouds, while the ——.——.——.—— line represents an IFR flight which avoids actual IFR conditions.

In every case, the total fuel-consumption figures will include nine pounds (1.5 gallons) of fuel consumed during start-up, taxi and takeoff.

Takeoff

We're now ready for the first flight. We take off from Wichita and climb to 3,500 feet. We then level off and head first toward Hutchinson (28 nm at 296 degrees) and then to Dodge City (99 nm at 257 degrees) and from there to Garden City (34 nm at 266 degrees). The climb to that altitude will take only a few minutes during which we'll be traveling about three nm, consuming about three pounds (0.5 gallons) of fuel, assuming a rate of climb of approximately 665 fpm.

Now, level at 3,500 feet. There is nothing wrong with flying westward at an odd-thousand-plus-500-foot altitude because we are less than 3,000 feet agl. The time to Garden City will add up to one hour and 12 minutes with a power setting of 2400 rpm and 20.2 inches of manifold pressure (55 percent of power). This produces a TAS of 140 knots and a ground speed of 130 knots—figuring that the headwind component will amount to approximately 10 knots. At Garden City, we climb 1,000 feet to 4,500 feet and continue on to Lamar. We then climb another 1,000 feet to 5,500 (still within less than 3,000 feet agl) for the stretch from there to Pueblo. The leg from Garden City to Lamar, not counting the time to climb, will take 43 minutes. The leg from there to Pueblo will take 35 minutes.

By adding the various climb times, we have now been airborne for two hours and 35 minutes and have used a total of 160.7 pounds (26.8 gallons) of fuel.

Climb

Now we have to climb in order to get over the mountains. While the mountain tops rise to above 14,000 feet, it is possible to cross through the passes at about 12,000 feet. That is the altitude we have selected for the next leg. The climb from 5,500 (at Pueblo) to 12,000 feet at a climb rate of about 540 fpm will take about 12 minutes. During that time we will cover a little under 20 nm horizontally and we'll be burning just under 18 pounds (three gallons) of fuel. Then, level at 12,000 feet. It will take 55 minutes to get to Gunnison, give or take a few minutes which may be involved in making minor detours in order to take advantage of the available mountain passes. From Gunnison we set up a rate of descent of about 166 fpm. This will get us to the altitude of the Grand Junction airport in 47 minutes.

Totals for the entire trip from liftoff to touchdown amount to four hours and 26 minutes in terms of time and 260.8 pounds (43.47 gallons) in terms of fuel burned.

An Alternative

Now let's look at the alternate way of flying that same trip which is also VFR. We take off from Wichita and climb directly up to 12,500 feet for a total of 11,168 feet. At an average rate of climb of approximately 500 fpm, this will take a little over 22 minutes. During that time we'll be consuming about 27 pounds (4.5 gallons) of fuel. Now, level at 12,500 feet, we fly at a power setting of 2400 rpm and 19.6 inches of manifold pressure (56 percent of power), producing a TAS of 151 knots. But we are flying into a 30-knot headwind. Therefore, our ground speed is reduced to 121 knots. The flight from the time we reach our altitude to the Gunnison VOR will take approximately three hours and 46 minutes and use 211 pounds (35.2 gallons) of fuel. We then set up a gradual descent toward Grand Junction that is similar to the one described in the previous flight.

The totals for this flight add up to four hours and 50 minutes flying time at a fuel consumption of 369 pounds (61.5 gallons). In other words, it took 24 minutes longer and used 18 additional gallons of fuel.

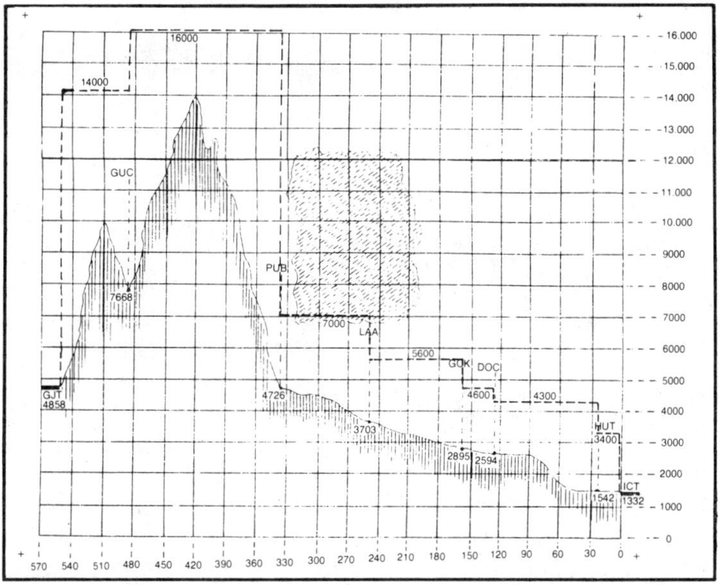

Fig. 9-3. The MEAs (minimum enroute altitude for IFR flight) between Wichita and Grand Junction.

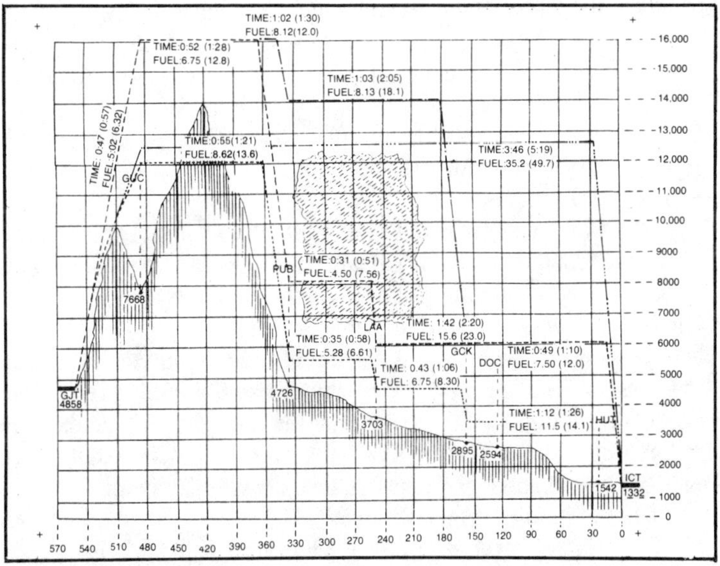

Fig. 9-4. The 201 and Arrow IV pilots have a number of choices:.......= VFR at the lowest possible altitudes;___.... ___....___....= VFR at 12,500 feet;----IFR including approximately 100 nm in the clouds;___.___ .___.___.___.=IFR avoiding actual IFR conditions. The time and fuel figures without parentheses are for the 201. The ones in parentheses are for the Arrow IV.

Trip Profiles

Now we'll examine the two suggested IFR trip profiles. In the first instance, we would request an initial altitude of 6,000 feet to Lamar, then from Lamar to Pueblo we'd ask for 8,000 feet, and from there to Gunnison the lowest MEA is 16,000 feet. After that, we'll assume that we cancel IFR and set up a gradual descent all the way to Grand Junction.

The climb from Wichita to 6,000 feet (a 4,668 altitude differential) will take about 8.5 minutes at a rate of climb of close to 800 fpm and use 14 pounds (2.3 gallons) of fuel. Level at 6,000 feet, we now cruise at a power setting of 2400 rpm and 22.8 inches of manifold pressure (65 percent of power). Allowing for the 15-knot headwind, it will take us one hour and 42 minutes at a TAS of 154 knots to arrive at Lamar. We now climb into the clouds to a new 8,000-foot altitude and then fly level at that altitude to Pueblo where we'll again be in VFR conditions. The time to climb will be about five minutes and the level flight will take 31 minutes. By now, we have been airborne for about two hours and 27 minutes and we have used 140.6 pounds (23.4 gallons) of fuel.

At Pueblo, we climb to 16,000 feet which, at approximately 400 fpm, will take 20 minutes and use 22 pounds (3.7 gallons) of fuel. Level at 16,000 feet, it will take 52 minutes to get to the Gunnison VOR, and then, at a slow rate of descent (about 235 fpm) it will take another 47 minutes to get to Grand Junction.

Totals for the trip, from liftoff to touchdown, have amounted to four hours and 23 minutes at a cost in terms of fuel of 232.3 pounds (38.7 gallons).

The last variation on the theme calls for an initial climbout to that same 6,000 feet. Then, at Garden City, an additional climb to 14,000 feet. This is followed by a climb to 16,000 feet at Pueblo. The balance of the trip is the same as before.

The initial climb will take just under six minutes at 800 fpm and burn 14 pounds (2.3 gallons) of fuel. From there the leg to the Garden City VOR will take 49 minutes and use 45 pounds (7.5 gallons) of fuel. At Garden City, the climb to 14,000 feet at an average rate of about 400 fpm will take roughly 20 minutes, cover 23 nm horizontally and burn 21 pounds (3.5 gallons) of fuel. Level at 14,000 with a power setting of 2400 rpm and 18.2 inches of manifold pressure (52 percent of power) to achieve a TAS of 147 knots. In view of the prevailing headwind, this provides us with a groundspeed of 112 knots. After flying level at that altitude for one hour and three minutes we will arrive over the Pueblo VOR where

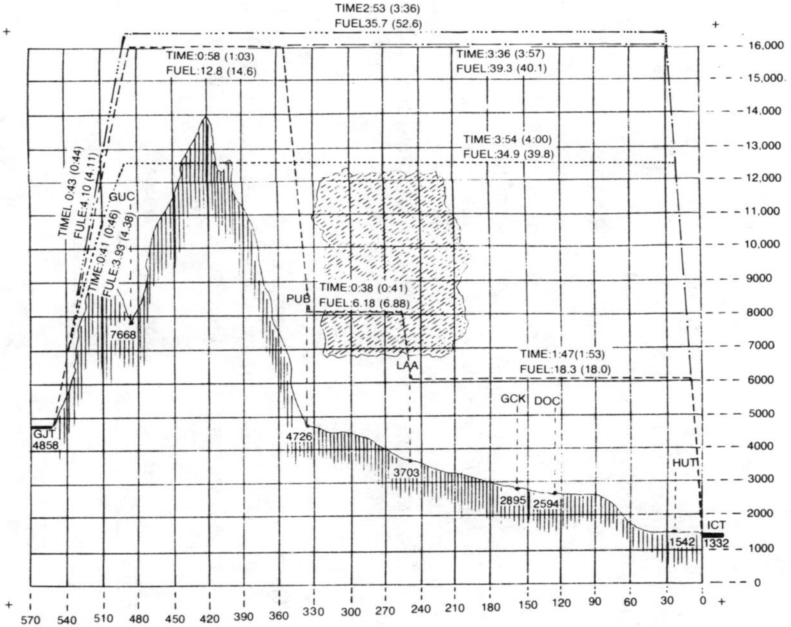

Fig. 9-5. These are the four possible trip profiles for the Turbo 231 and the Piper Turbo Arrow IV: -----= IFR at the lowest altitudes permissable, including approximately 100 nm in the clouds; ——.——.—— = IFR at a high altitude all the way; ....= VFR at 12,500 feet; ——....——....—— = VFR at an altitude at which the turbocharger comes into its own. Time and fuel figures without parentheses are for the Turbo 231. The figures in parentheses are for the Turbo Arrow IV.

Table 9-1. Time and Fuel Comparisons.

	Total time	Total fuel
Trip #1:	4:26	43.47 gal.
Trip #2:	4:50	61.5 gal.
Trip #3:	4:23	38.7 gal.
Trip #4:	4:17	35.6 gal.

we'll start the climb to 16,000 feet. We then continue at that altitude for one hour and two minutes to reach the Gunnison VOR where we'll again cancel IFR and descend for 47 minutes at approximately 235 fpm to arrive at Grand Junction at the desired altitude.

In this case, the totals amount to four hours and 17 minutes at a fuel consumption of 213. 6 pounds (35.6 gallons). To compare the four alternatives, see Table 9-1 for the totals in terms of time and fuel.

The Turbo 231

Now let's cover the same distance in the Mooney Turbo 231. Again we have picked two IFR and two VFR profiles. The first IFR profile is the same as was flown with the 201. It is shown in Fig. 9-5 as a ------ line. Using these altitudes, allowing for climb, winds aloft and descent, plus 10 pounds (1.7 gallons) of fuel for start-up, taxi and takeoff, the total time comes to four hours and six minutes while using 262.8 pounds (43.08 gallons) of fuel.

The second IFR profile calls for a climb directly up to 16,000 feet and remaining at that altitude until Gunnison. Then there is a gradual descent to the destination. Here to total time adds up to four hours and 19 minutes while using an economical cruise setting at altitude. The fuel burned totals 270.6 pounds (45.1 gallons). See the —.–—.——. line on the chart.

The first VFR profile calls for a climb to 12,500 feet, a cruise at that altitude to Gunnison and then the descent. Here the time total is four hours and 35 minutes while accounting for a fuel

Table 9-2. Time and Fuel Comparisons.

	Total time	Total fuel
Trip #1:	4:06	43.08 gal.
Trip #2:	4:19	45.1 gal.
Trip #3:	4:35	40.53 gal.
Trip #4:	3:36	41.5 gal.

Table 9-3. Aspirated Arrow IV Results.

	Total time	Total fuel
Trip #1	5:32	50.66 gal.
Trip #2	5:40	50.72 gal.
Trip #3	5:38	49.11 gal.
Trip #4	5:57	56.2 gal.

consumption of 243.18 pounds (40.53 gallons). See the ..... line on the chart.

The second VFR profile (which could just as well be flown IFR) calls for a maximum-performance climb all the way to 16,500 feet, then full-throttle cruise at that altitude and the usual descent. Here the time is reduced to three hours and 36 minutes with a total fuel cost of 249 pounds (41.5 gallons). See the ——....——....——line of the chart. See Table 9-2 for how these four trip profiles line up.

It is interesting to note that despite the considerable 35-knot headwinds at altitude, the 231 saves so much time by being flown at maximum power (2700 rpm and 28.5 inches of manifold pressure) that even at the much higher fuel flow it ends up using barely one gallon more than the most economical configuration provides. Anywhere from a one-half hour to nearly an hour is saved over the other trip profiles.

Table 9-4. Turbo Arrow IV Results.

	Total time	Total fuel
Trip #1	4:34	45.29 gal.
Trip #2	4:41	60.51 ga.
Trip #3	4:46	45.86 gal.
Trip #4	4:20	58.41 gal.

In order to compare the two Mooney aircraft with the two Arrows, we will fly each airplane using the same altitudes and trip profiles. The time and fuel figures for the Arrows are shown on the charts in parenthesis. For the normally aspirated Arrow IV, the results are shown in Table 9-3. For the Turbo Arrow IV, the results are shown in Table 9-4.

The advantage of the Turbo 231, which is shown in Fig. 9-5, holds up when an actual comparison is run. In fact, when that aircraft is flown somewhere close to its optimum altitude, it might as well be flown at a high power setting. The time saved easily makes up for the increase in fuel flow.

Chapter 10

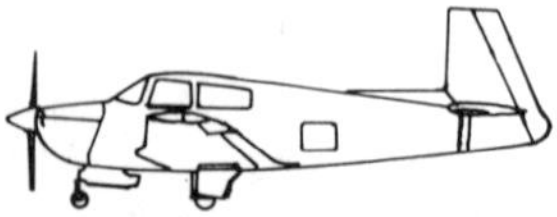

Mooney's Owner's Manuals

It is the consensus among aviation people that Mooney, in preparing the owner's manuals for the 201 and 231, did one of the best and most thorough jobs ever done for any single-engine aircraft. This is undoubtedly true in that these manuals do contain all the possible information that an owner/pilot might want. A problem with both of them, but especially the manual for the 231, is that specific information is extremely hard to find. Many of the charts are so complicated and the printing on them is in such small type, that anyone with less than perfect eyesight would need a magnifying glass to figure out what it says. Worse yet, trying to use these charts in flight would appear to be a virtual impossibility. Since the very first page states: *This handbook includes the material required to be furnished to the FAR part 3 and must kept in the airplane at all times*, this tends to present a problem.

Handbook Data For The 201

Let's start by looking at the handbook for the 201. It starts off with a table of contents which lists the 10 sections into which the material is divided. These are: General—Limitations—Emergency Procedures—Normal Procedures—Performance—Weight and Balance—Airplane & System Descriptions—Handling, Service and Maintenance—Supplemental Data—Safety Information.

What is missing, although it would represent a huge improvement in the usefulness of the handbook, is an index which would list every piece of information and parameter contained in the book. Granted, each section has its own table of contents, but it is often difficult to locate a specific piece of needed information. For instance, if you wanted to find the *useful-load* figure, you might logically look under Weights in the Limitations section. However, it isn't there. It does show up under General/Standard Airplane Weights.

The 201 *Pilot's Operating Handbook and Airplane Flight Manual* consists of a total of 180 pages (not counting the section dividers). For obvious reasons there isn't room here to reprint the entire manual. Rather, we'll take a close look at the information and at the various charts and illustrations (supplied courtesy of Mooney) which represent the more important data.

Section I—General

First there is a three-view of the aircraft, giving its configuration and dimensions (Fig. 10-1). Then there are two pages of descriptive data. Covered are such things as the type and size of the landing gear, engine parameters, propeller, total and usable fuel, oil, maximum certificated and standard airplane weights, baggage compartment dimensions, and the figures pertaining to wing- and power loading. This is followed by five or so pages of explanations of the various terms and appreciations which are used throughout. For some reason or other, they are not in alphabetical order. This makes finding a specific term difficult.

Section II—Limitations

This section starts with a table of airspeed limitations that are given in statute miles (mph). The most important are V_{ne} (225), V_{no} (200) and V_{lo} (150). The meanings are explained for the colored bands on the altimeter, power-plant limitations and instrument markings, weight and center-of-gravity limits, flight load factors, plus various instrument markings and standard placards.

Section III-Emergency Procedures

This is the section which, although identified in red in order to be found quickly, most of us never bother to read. Perhaps this is because we're habitually convinced that emergencies only happen to the other guy. It starts off with an explanation of the annunciator panel warning lights, and then goes on to explain what to do in the event of engine fire on the ground or in the air, electrical fire, and engine failure, before, during or after takeoff. Under this heading, it includes a relatively simple chart which shows the best glide distance which can be achieved under given conditions (Fig. 10-

2). It continues with advice relative to a power-off landing, an unlatched door in flight, alternator power loss, flight in icing conditions, alternate static source, and the failure of the gear to retract after takeoff—including information on how to get it down again.

Section IV—Normal Procedures

Here we start off with the typical description of the recommended preflight-inspection procedure, prestarting check, starting procedure, flooded-engine clearing and pre-takeoff checks. Then there are the instructions for normal takeoff, climb, cruise, and descent, plus go around, landing, taxi, shutdown and finally securing the aircraft.

Section V—Performance

This is the section which most of us are likely to use more often than any of the others. It contains the kind of information that is needed during virtually every flight if you want to achieve maximum efficiency coupled with the greatest possible economy. Virtually all of this section consists of charts and tables with relatively little description. A pilot would do well, indeed, to study these charts in the comfort of home on a rainy day so that, when the information is needed in flight, he understands what he is looking at. It should be pointed out that, while airspeed limitations in the earlier section were given in mph, in this section all speeds are in knots. Temperatures are given in centigrade.

The charts have been reproduced in the consecutive order in which they appear in the handbook. See Figs. 10-3 through 10-11 and Tables 10-1 through 10-23.

Section VI—Weight And Balance

This starts with an explanation of the authorized airplane-weighing procedures, followed by an explanatory illustration (Fig. 10-12) and a weight-and-balance record chart designed to be completed by the pilot (Fig. 10-13). Then there is an explanation of how to load the aircraft, a problem form and loading-computation chart (Fig. 10-14), a chart showing the center-of-gravity moment envelope and a center-of-gravity limits chart (Fig. 10-15). This is followed by a list of the weights and moment arm for all optional equipment which is approved for the 201 (M20J). There are 11 pages of charts showing how to enter the data into the weight-and-balance record. Although this section includes all the needed information and instructions, the weight-and-balance computer developed and used by Piper is a great deal easier to use.

Section VII—Airplane & Systems Description

Here the handbook has information on the airframe, the power plant, the ignition system, the fuel system, engine cooling, vacuum system and the propeller. There are graphic illustrations of the engine air-induction system (Fig. 10-16) and the fuel system (Fig. 10-17). This is followed by a chapter headed "Flight Panel and Controls Familiarization" which explains the functions of all switches, basic instruments and controls (Figs. 10-18 through 10-21), plus a diagram of the electrical power setup (Fig. 10-22) and the positioning of the circuit breakers (Fig. 10-23). And then there is a description of the heating and ventilation system (Fig. 10-24) and some advice relative to seatbelts and shoulder harnesses.

Section VIII—Handling, Servicing & Maintenance

This section covers ground handling (towing, tiedown, jacking); servicing (refueling, engine lubrication, air filters, gear and tires, battery, and hydraulic brake care); and maintenance (propeller care, exterior care, interior care). See Fig. 10-25 through 10-28.

Section IX—Supplemental Data

This section is blank.

Section X—Safety Information

This section explains what to do and what not to do in general terms and about sources of additional information. Covered are Federal Aviation Regulations, advisories, NOTAMs and the Airman's Information Manual, flight planning, turbulence, mountain flying, stalls, spins, wake turbulence, vertigo, etc. And there is also a section on medical facts for pilots. Most of this material is too basic to be of much interest to a pilot with any appreciable degree of experience.

Handbook Data For The Turbo 231

On the whole, this handbook is fairly easy to use and most of the frequently needed information is presented in a manner which is easy to understand. This is not the case with the handbook for the

Turbo 231. It too is divided into 10 sections. It starts with the three-view (Fig. 10-29) in Section I, a list of the various limitations and placards in Section II, emergency procedures including a visual representation of the maximum glide distance (Fig. 10-30) in Section III and a description of all the normal operating procedures in Section IV.

But now comes the rub. Section V—Performance, starts off the applicable conversion charts (Figs. 10-31 and 10-32). These are followed by three airspeed calibration charts (Figs. 10-33, 10-34 and 10-35) which relate indicated airspeed to calibrated airspeed under a variety of gear, flap, cowlflap and static-source conditions. Frankly, during the day-to-day operation of the aircraft, they are of little importance. The next three pages contain charts showing methods for correcting altimeter errors under various gear, flap and static-source conditions (Figs. 10-36, 10-37 and 10-38). These charts, too, will probably be ignored by most pilots most of the time.

Stall Speeds

The next page contains information which is of greater importance. Here we have a table of stall speeds (in mph and knots) at various angles of bank from zero to 60 degrees (with and without flaps). This is the sort of information which, if memorized by the pilot, could help keep him out of trouble at times when wind conditions call for steep turns at low altitudes (Fig. 10-39).

Takeoff Parameters

The next two charts get a bit more complicated. They represent the takeoff ground-roll distance and the distance to clear a 50-foot obstacle under varying conditions of temperature, wind and aircraft weight (Figs. 10-40 and 10-41). It takes a bit of study to figure out the meanings of the various lines and curves, but since takeoff parameters are usually studied on the ground, it can be assumed that there is ample time to familiarize yourself with the data before attempting a minimum-distance takeoff.

Rate Of Climb

Figure 10-42 is a chart entitled Rate of Climb. This, too, assumes variations in density altitude and aircraft weight. Since circumstances under which the rate of climb are important, it might best be studied on the ground. The same holds true for the next chart which depicts the time, distance covered and fuel used during climb to altitudes from zero to 24,000 feet (Fig. 10-43). The chart provides this information for two different aircraft weights, 2,900 pounds and 2,300 pounds, and it wouldn't be too difficult to interpolate the figures for other aircraft weights.

Power Schedules

The next two charts are cruise power schedules. They show the different rpm and manifold pressue settings to achieve a given percent of power at altitudes from sea level to 24,000 feet. They also show the fuel flow in gph for the different rpm/MP combinations (Table 10-24 and Table 10-25). Since these charts are likely to be used more often in flight than on the ground, it is too bad that they are printed in such small type that reading them in the cockpit at night or under conditions of a degree of turbulence would be quite difficult.

The next chart is headed Speed Power vs. Altitude (Fig. 10-44). This is one of those charts that requires drawing lines with a ruler to figure out the true airspeed which will be achieved at different percentages of power under differing weight, pressure altitude and density altitude condition. Again, to use this chart in flight might prove to be nearly impossible.

Range

We now come to five charts. These show the maximum range of the aircraft (allowing for warmup, taxi, takeoff, climb, and 45-minute reserve) when flying at different altitudes while using either best economy or best power at varying percent-of-power settings. These charts (Figs. 10-45 through 10-49), and virtually all those that follow in the section, are probably easily understood by engineers and professional corporate pilots. But since many 231s are likely to be flown by pilots whose main interest is their business or profession rather than the acquistion of intimate knowledge of the finer points of aircraft performance, it would seem that a more easy-to-comprehend representation of these data would be a definite improvement.

The next five charts (Figs. 10-50 through 10-54) show the maximum endurance in terms of hours under different percent-of-power conditions. This is assuming either best power or best economy. Best power assumes flying at 125 degrees on the rich side of peak, while best economy involves flying at peak turbine-intake-temperature.

The next two charts are similar to those showing takeoff distances. However, these refer to landing over a 50-foot obstacle and the landing ground roll under a variety of conditions (Figs. 10-55 and 10-56).

What follows are two pages of detailed instructions on the subject of how to use the nautical-miles-per-gallon charts which follow. Quite frankly, without reading these instructions, it would be pretty difficult to understand what the lines and curves represent. Preceding the nm-per-gallon charts is a power schedule (Table 10-26) which shows the recommended rpm/MP combinations for differing percentages of power at varying altitudes under standard-temperature conditions. This is then followed by the nm-per-gallon charts for all altitudes from sea level to 24,000 feet (Figs. 10-57 through 10-69). These seem to indicate that by flying at a low percent of power at a very high altitude (24,000 feet) it is possible to achieve an astounding 30 nm per gallon.

That takes care of the performance section. It contains a world of important and quite reliable information for those who are willing to take the time to carefully study all those charts and relate the information gained from one to the information contained in others. Personally, if I owned a 231, I would probably take the handbook home and prepare some simple-to-read-in-flight charts. The charts would show the data which I'd most likely want to use with any frequency during normal operation.

The rest of the handbook is fairly routine. Section VI Weight and Balance, uses some easy-to-comprehend illustrations (Figs. 10-70 through 10-75), to explain the weight-and-balance problems which are a bit more complicated and critical than they are in the 201.

Section VII—Systems covers the airframe, flight-control systems, cockpit instruments (Figs. 10-76, 10-77 and 10-78), ground control, landing gear, the engine—with illustrations of the oil system, air-induction system, the turbocharger, the fuel-injection system and the fuel system in general (Figs. 10-79 through 10-83) and the electrical system, circuit-breaker panel and the heating/cooling system (Figs. 10-84, 10-85 and 10-86).

Section VIII—Handling, Servicing and Maintenance, is comparable to that section for the 201. Section IX—Supplemental data, is blank. Section X—Safety Tips, considering the sophistication of the performance section and the airplane, is not much more sophisticated than that section in the 201 handbook. The total number of pages in the 231 handbook is 228.

Illustrations and tables referenced in this chapter are on pages 151-222.

Chapter 11

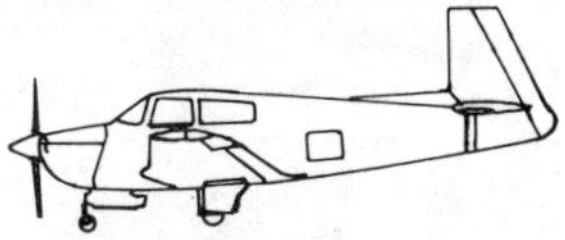

Turbo 231 Sets A World Record

At 6:49 a.m. on January 7, 1980, a 27-year-old flight instructor and charter pilot named Alan Gerharter lifted a Mooney 231 (Fig. 11-1) off the concrete at San Francisco International Airport with Washington D.C. as his destination. The purpose of the flight was to attempt to break a record set by a Cessna Turbo Centurion in January of 1977.

Flight Preparation

In preparation for the flight, he plotted a great-circle route from SFO to DCA and established 21 waypoints which he planned to use with the on-board RNAV equipment. Since the aircraft he was planning to fly did not have a built-in oxygen system, he replaced the right front seat with two portable oxygen tanks and then installed two 25-gallon auxiliary fuel tanks in the backseat area of the airplane. This increased the amount of usable fuel to something in the neighborhood of 122 gallons.

A fairly strong jet stream was forecast at altitude so he climbed straight up to 25,000 feet where the weather was clear except for a few higher cloud tops in the area around Denver. As seems to happen with depressing regularity during such record attempts, something started to go wrong within less than two hours after takeoff. The alternator began to act up for one reason or another, causing the DME to cut out intermittantly. This made the RNAV equipment useless for all practical purposes. As a result, he was forced to fly the rest of the trip using his VHF nav receivers and his ADF.

Fig. 11-1. The Turbo 231.

The New Record

At that altitude, Gerharter used a powersetting combination of 2700 rpm and 28 inches of manifold pressure which produces approximately 75 percent of power. At that setting and with the jet stream pushing him along at an additional 100-plus knots, he ended up with a ground speed of just under 300 knots. By the time he landed at Washington National Airport some eight hours, four minutes and 25 seconds after liftoff, he had burned a total of 618 pounds (103 gallons) of fuel. Even at $2 per gallon this made the trip a lot cheaper than an airline ticket. The Turbo Centurion's record was beaten by three hours, 56 minutes and 37 seconds at a fuel cost of less then two thirds (in terms of quantity) of that used by the Cessna.

Gerharter is thinking of repeating the trip, this time using a miserly power setting of 35 percent, to find out for himself how economically such a flight could be made and at what cost in terms of time.

Chapter 12

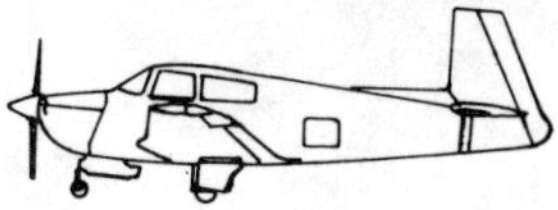

Maintenance Characteristics

There is one man who knows more about what makes a Mooney airplane tick than probably anyone else anywhere in the world. Charley Dugosh owns the maintenance facility at the Kerrville airport and Mooney owners come to him from every part of the country and from as far as Canada and Mexico to have their maintenance work and repairs done by him.

Dugosh says that the advantage he has over virtually anyone else who works on Mooneys is that he was there during the early stages of the testing procedure. "That kind of experience teaches you a lot," he says. "You find out why you need certain size tires and why the right air pressure in your tires is important."

He was part of every phase of the early testing. The original wing was designed for a 2,000-pound airplane. Then it was decided to go to 2,200 pounds and the fabric cover on the rear portion of the wing was replaced by plywood. During the static testing, Al Mooney decided to increase the gross weight to 2,450 pounds. Therefore, the 2,200-pound wing was tested for 2,450 pounds. Dugosh was one of the men holding the end of the wing, upside down, while it was gradually loaded down with sand. He says that as best as he can remember it bent 23 inches before it broke. Subsequently they beefed it up some more to allow for the extra weight. Those years with Mooney, he says, gave him an advantage over other mechanics who never get a chance to see how an airplane is actually built.

Wing Strength

The Mooney wing (Fig. 12-1) on today's all-metal Mooneys is famous for its fantastic strength. It's all one piece except for the

Fig. 12-1. The Mooney wing is one continuous structure except for the wing-tip panels (courtesy of Mooney).

wing-tip sections. "We've had them hit trees, bend wing tips," Dugosh recollects, "and all you do is strap a 2 × 4 to the wing and fly it away. It's so strong, we had a doctor who got into the overcase dive out and then pull out at what must have been excessive speed. We figured that he must have exceeded eight g's. It loosened hundreds of rivets, but we checked the dihedral and everything and couldn't find any distortion. And there was another guy, he flew a Statesman. He pulled out of a dive so hard that the passengers blacked out and the gear pulled out and bent the bell cranks. But nothing happened to the wing."

What actually happens is that the wing will bend some, but it's so stiff that it actually tries to shear the rivets. Even with as many as 200 rivets loose, it continues to remain safe. According to Dugosh, there have been only two incidents of metal Mooneys actually losing a wing in flight. In each case, it was estimated that the aircraft was exceeding 300 mph.

One serious problem that has faced Dugosh repeatedly is that when there is trouble somewhere out in the boondocks, people don't seem to know how to get the wings off the airplane. There are cases of the use of cutting torches to simply cut the wings off. Other attempts were made to chisel them apart. Quite obviously, the task of subsequently rebuilding what is left of the airplane becomes unnecessarily difficult.

Fig. 12-2. The one-piece wing is bolted to the fuselage by 14 bolts.

Actually, taking the wing off the airplane is quite simple. Says Dugosh, "I used to disassemble one in one day all by myself." The wing is bolted to the fuselage (Fig. 12-2), with 14 bolts and once those have been removed you can lift the fuselage right off the wing which, after all, is one piece on which the fuselage rests. And, once separated, the two components can easily be loaded on a trailer or even a pickup and carted away.

The Tubular Fuselage

The outer skin of the fuselage is built around a welded steel-tube, roll-cage (Fig. 12-3) which has proved to be the major reason for the Mooney's exceptional crash-worthiness. Explains Dugosh: "That tubular fuselage has saved more lives. There was a guy who ran out of gas at night and hit the trees. From the place at which he hit the top of one tree to where he finally came to rest at the trunk of another was a distance of about 20 feet. He got up and walked away with one tiny scratch on his head."

What worries Dugosh is that he believes that eventually that tubular structure may be eliminated because of the production cost associated with it. Its construction involves a lot of hand work and manhours. It is, without question, a major factor in the safety of that airplane. "Unless you dive it straight into the ground, you're going to survive, and that's why Al (Mooney) put it in there in the first place. I've never seen one of those cages pull apart."

The Nosewheel

Towing a Mooney is something that has to be done with an understanding of the steering limits of the nosewheel. Towing the airplane with a tractor is likely to bend the truss. The nosewheel can only be turned just so far. Any additional turn will result in the need to replace that truss which, today, costs $333.76. "We have replaced four of them in the last month alone. It's strictly from towing. There's no way you can damage it by pulling the airplane by hand."

Rigging And Landing Gear

Another unique feature is associated with the rigging of the landing gear. It uses push-pull tubes, two in the nose gear (Fig. 12-4) and one each in the main gears. Whenever anything is done to the gear, it has got to be re-rigged strictly strictly by the book. It is a symmetrical system and requires going back and forth, and back and forth to get it right.

Wood-Wings

There is still a reasonable number of the old wood-wing Mooneys in service. Basically, relative to its weight, wood is

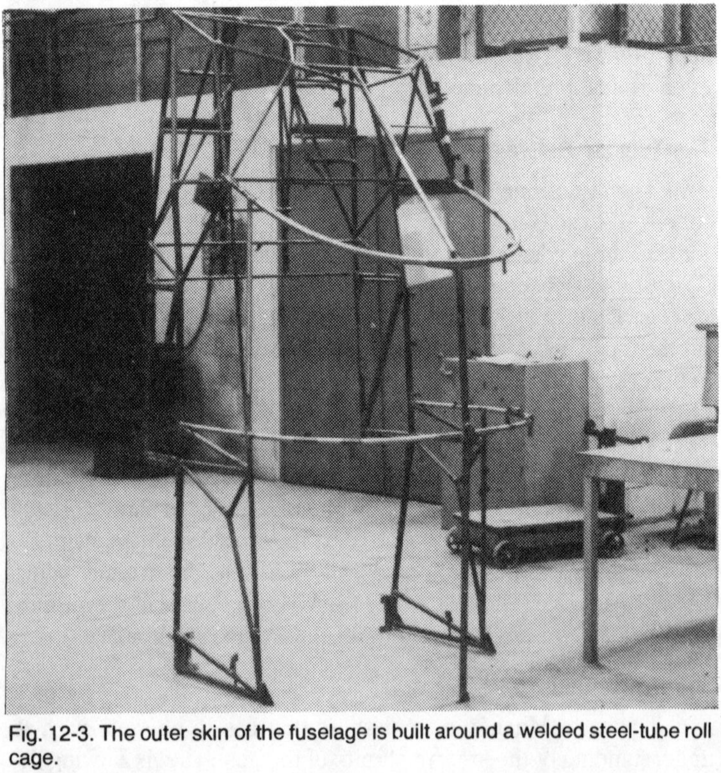

Fig. 12-3. The outer skin of the fuselage is built around a welded steel-tube roll cage.

actually stronger than metal and the wood-wings are a couple of hundred pounds lighter than the all-metal airplanes. But wood needs a lot of care. The biggest problem is moisture—either too much or too little—usually too much. If the airplane is always hangared and kept clean, trouble is not likely to develop. The wings have to be inspected every six months. At that time, the wheel-well and all the drain holes must be checked. Then, at each annual inspection, the fairings are pulled and resealed around the wing root and the fuselage to prevent water from seeping in there.

The wing itself is covered with fabric and painted with dope. A sure way to invite eventual trouble is to paint these wings with enamel. Enamel doesn't contract and expand at the same rate as fabric and dope. This causes cracks to develop. The only type of finish that can safely be used is dope. This simplifies the task of resealing any cracks which do develop.

Another important factor is the direction of the grain of the plywood. The airplane was built with the plywood in a 45-degree grain, but too many people who are called upon to perform repair work on these wings don't seem to know enough about them. They will put in plywood at 90-degree grain. Either they do not know or do not care that this reduces the structural strength.

There is an instance on record which illustrates that too little moisture can be as detrimental as too much moisture. There was a wood-wing from Arizona which was brought to Dugosh for inspection. It was so dry that every single rib had shrunk away from the main spar. They all had to be replaced and reglued.

As a general rule, Dugosh feels that the trouble with too many maintenance shops is that they would much prefer to work on large airplanes, "because they can hide the bills. My idea has always been that if I work on a little airplane for a guy, one day he may own a fleet, and if you treat him right all along, he'll stick with you."

Auxiliary Tanks

A large number of Mooneys have been sold to buyers overseas. Dugosh has installed auxiliary tanks for Mooneys which were to be flown to Europe, Asia, Australia—all over. "I've got one sitting out there right now that's supposed to go to South Africa and we've got to tank it." He has developed a specially built tank that fits in the space of the copilot's seat, holding 58 gallons, and another designed to fit in the back of the airplane that holds 66 gallons.

Those tanks are considered expendable. When the aircraft arrives at its destination, they are thrown away. Many years ago,

Fig. 12-4. Another unique feature is the kind of rigging employed in the nose gear.

the company once tried to ship their aircraft demated and crated, but that was never repeated. The trouble is that once an airplane is taken apart and then put back together again, it simply doesn't seem to fly quite right if the people doing the work don't know exactly how to do it with the necessary precision. By ferrying them, all the bugs are worked out of the airplane right away. Granted, they've got a few extra hours on them, but at least they are ready to go to work. Today, Dugosh reports, a lot of 201s are being sent to Europe—especially Germany.

Those airplanes are proving increasingly popular largely as the result of the fuel situation not only in this country but elsewhere. It cannot be argued that, relative to fuel consumption, they are the most economical airplanes anywhere. Basically, of course, the current models are exactly the same as the early all-metal Mooneys. The painstaking cleanup work has made them faster and better looking, and there have been many minor changes to reduce drag and improve appearance. But the basic structure has remained intact.

Does the turbocharger in the 231 result in an increase in maintenance problems and cost? Dugosh says: "In fact there is no appreciable difference in overhaul time if the pilot uses his turbocharged engine correctly, according to the book. Oh, it's a little harder to work on, and, of course, there is the extra maintenance of the turbocharger itself, but that really doesn't amount to much."

All in all, it's nice to know that there is a place where the people who work on our airplanes love them with a passion and who, therefore, can be relied upon to take great pains to see that any airplane leaving the shop is in perfect operating condition.

Chapter 13

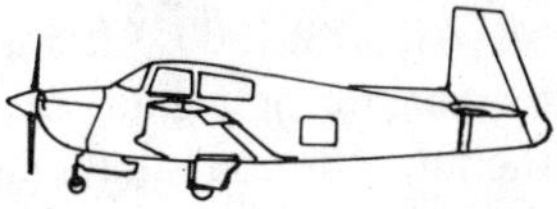

Not Every Mooney Is A Mooney

After Mooney Aircraft Inc. had successfully established a distributor/dealer network which was selling various models, first the Mark 20 and the Special 21 and then the Ranger, Chaparral, Statesman and Executive, the dealers kept applying an increasing amount of pressure on the company to extend the product to a more complete line of aircraft. They wanted a simple 2-seater which could be used for training, a relatively easy-to-fly, fixed-gear 4-seater, and a twin of some kind.

The reasoning was perfectly sound. As long as flight students had no choice but to take their lessons in a Cherokee 140 or Cessna 150, they were likely to develop a degree of product loyalty. This made it that much more difficult to eventually sell them a Mooney. Conversely, Mooney owners who wanted to trade up had no choice but to look at, and eventually buy, the product of another manufacturer.

The Mooney Master

Of the various dealer demands, the one that was easiest to satisfy was the one for a fixed gear 4-seater. All that was needed was to take a standard Mooney and nail down the gear. That was done and the result was the Mooney Master which, in the long run, turned out to excite neither the company nor its customers. Production of the Master never reached any meaningful proportions.

The Aircoupe

In order to add a trainer to its line, Mooney bought Alon Inc. They were manufacturers of what had started out as an Ercoupe and was now marketed as the Aircoupe. The original Ercoupe, designed and manufactured by Engineering Research Corp. (Erco), saw the light of day in 1937 as a twin-tailed, non-spinnable, rudder-pedalless metal and fabric 2-seater. Powered over the years by a variety of engines, it was a very popular airplane in its day. In the mid-1950s, production rights were acquired by the Forney Co. which rechristened the airplane Aircoupe and manufactured it until 1960. After a five-year hiatus, Alon Inc. purchased the manufacturing rights. After installing rudder pedals and changing the aircraft to all-aluminum, it manufactured and sold the airplane as the Alon Aircoupe. That lasted until 1968 when Mooney bought the rights. In order to make it look more like a Mooney, the twin-rudder tail was elimianted and replaced by a typical canted-forward Mooney tail. Stall strips were added to the leading edges of the wings to make the stall characteristics more in line with those required by flight instructors. Mooney renamed the airplane the M-10 Cadet and now they had a trainer.

Of course, calling it the M-10 was patent nonsense. Al Mooney had by then designed some 20-odd airplanes, each identified as M-1, M-2 and so on. Of course, this included an M-10 (a Monocoupe, designed in 1935). Though some of the Cadets are still flying today, the airplane was relatively unsuccessful and production was halted in 1970. Of course, it is possible that if Mooney's money troubles hadn't interfered, it might eventually Have been developed into a successful training machine.

So much for the bottom end of the projected Mooney line of airplanes. The obvious choice for an additional model at the top end was a light twin. In fact, the Mooney people at one point did hang two engines on what was, for all practical purposes, a somewhat beefed-up standard Mooney. But that program was quickly scuttled when Ralph Harmon decided that what general aviation needed was an affordable single-engine turbocharged and pressurized aircraft. He convinced Mooney management that he could design and build such an aircraft out of cash flow without the need for massive additional investment and that the final airplane could be sold for $29,000.

The Mooney Mustang

Enter the M-22 Mooney Mustang. The M-22 has nothing actually to do with Al Mooney. His M-22 was a jet-powered VTOL military aircraft designed by him for Lockheed. The Mustang cost four times as much to develop as had originally been anticipated. It not only didn't look like a Mooney, but it actually was not an attractive airplane. It was difficult and expensive to build, but it did have some pretty sensational performance parameters. Cruise speed was 240 mph (209 knots) at altitude.

Meanwhile, the irrepressible Hal Rachel, Mooney's president, had been trumpeting to all who would listen that the airplane would be sold for $29,000 and that it would be such a fantastic airplane that competitors such as Cessna and Beech would be driven right out of business. If he had been quiet until he knew what kind of an airplane he actually had there, the final outcome might have been different. Instead, he was committed to deliver a given number of the pressurized Mustangs at the originally announced price of $29,000. However, it was costing somewhere in the neighborhood of $50,000 to build it. Today, of course, even $50,000 sounds like peanuts for a pressurized single. But in those days, it is doubtful that any single-engine airplane, no matter how fabulous, could have been sold for more than possibly $35,000 on the outside. The Mustang, another of the non-Mooney Mooneys, was the airplane that eventually drove the company into financial ruin.

The Mooney MU-2

In the meantime, there was another Mooney that wasn't a Mooney. It was the Mooney MU-2, the twin-engine turboprop that had been developed by Mitsubishi Heavy Industries in Japan and

Fig. 13-1. The MU-2 was assembled by Mooney in a newly built plant in San Angelo, Texas (courtesy of Mooney).

Fig. 13-2. Today's longer and quieter MU-2s are very successful in the corporate aviation field. The name, Mooney MU-2, was changed to Mitsubishi MU-2 after Mitsubishi severed connections with Mooney.

which was assembled by Mooney in San Angelo, Texas (Fig. 13-1). It was marketed, under the Mooney/Mitsubishi agreement, as a Mooney aircraft. The trouble was that initially they found that they couldn't give that airplane away for sour apples. It was noisy, not particularly comfortable, extremely difficult to fly and, when sitting on the ground, it kind of looked like a duck with a bad case of diarrhea. It took a while, too long for Mooney in its precarious financial condition, to improve the airplane to the point where it became acceptable in the U.S. corporate market. Success gradually started with the F model and has been growing ever since as the MU-2 has been made quieter, larger, more comfortable and easier to fly (Fig. 13-2).

Other Mooneys

The only true Mooney airplanes throughout all those years—not counting the single-seat M-18 Mite and its military counterpart, the M-19—were the M-20s. They were marketed under a variety of names. The original wood-wing Mooney 4-seat retractable aircraft (the M-20) was marketed as the Mark 20. Other models were the M-20A: the Mark 20A; M-20B: the Mark 21; M-20C: the Mark 21; M-20D: the Master; M-20E: the Super 21; the M-20F: the Executive and the M-20G: the Statesman.

When Republic Steel bought and revitalized the company, they decided to skip a few letters. Therefore, the M-20J is the 201 and the M-20K is the Turbo 231. One of these days, there will be another non-Mooney Mooney. It is temporarily called the M-X. This is going to be another pressurized single-engine airplane, this time designed by Roy Lopresti, with no relationship in any way, shape or form to the Mustang. Roy is shooting for a 300-mph cruising speed (261 knots). Only time will tell.

Chapter 14

What Next?

Any company that stands still does, in fact, go backwards. Therefore, it is only logical that the people at Mooney are in the process of making plans for the next airplane to join the lineup. While only sketchy information about that airplane is currently available, Roy Lopresti's enthusiasm with reference to it is catching:

"The next airplane is so good that I can hardly believe it. Oh, it's still about four years away. We just finished the preliminary design, and there are just literally thousands of individual parts to be designed (Fig. 14-1). It's a totally new airplane. We thought at one point that we might save the wing because the basic Mooney wing is so good, but it's really overdesigned. They used it years ago on the Mustang and that airplane had a gross of 3,600 pounds, so that wouldn't have been a problem from a stress point of view. But the wing is really older technology, and even though we knew we could use it and get away with it, we felt that the airplane, which is going to be a good new airplane, should not be penalized by something we did before. So the new airplane will be totally new, new wing, new airfoil, high wing loading, just a superb airplane."

The M-X

The subject of all this enthusiasm, currently referred to at the factory as the M-X, is a 6-seat single-engine pressurized airplane which, according to scattered reports which have appeared in *Flying* and other aviation publications, will be cruising at 300 mph. And if past experience is any critierion, it will be an airplane that is

Fig. 14-1. It takes literally thousands of parts, all individually designed, to build an airplane (courtesy of Mooney).

pretty to look at, easy to fly, miserly with its fuel, and fast, fast, fast.

According to bits and pieces of information which have filtered down to the aviation community, the airplane is likely to have an airfoil with increased lift distribution over the chord. This will permit the effective use of spoilers, much longer flaps and shorter, so-called, feeler ailerons. The cabin will be more spacious than we are used to on the earlier Mooneys. And despite the anticipated high cruising speed, the airplane will have to stall at just under 60 mph in order to comply with FAA regulations.

Pressurized Singles

While there can be no doubt that by the time the M-X (or-whatever its eventual name) is ready to take its place in the market, there will be other pressurized singles. A pressurized Bonanza and probably a pressurized version of the Piper Saratoga will likely be built. It would seem that the timing couldn't be better because, as we all know, the cost of fuel isn't going to go down and its availability is not likely to improve a great deal during the coming decade.

Chapter 15

The Mooney Brothers

Al Mooney (Fig. 15-1), the man who during his 45 years of active involvement in aviation designed a total of 23 airplanes, (not to mention several others which masquerade under his name but with the design of which he had nothing to do) was born in 1906 in Denver, Colorado. He went to various schools in Denver, Colorado and Pueblo, Colorado, and in Glendale, California and Burbank, California. He was graduated from South high school in Denver in 1924. He holds a private pilot certificate which was issued in 1929 with the number 5558.

He lives with his brother Art in a comfortable mobile home in Ingram, Texas, a few miles from Kerrville and the aircraft factory which bears his name—but which he has not visited in years. Here, surrounded by photographs and memorablia of his long and occasionally checkered career, he can look back to the time in June of 1925 when, as a cocky 19-year-old, he talked himself into a job as draftsman at the Alexander Aircraft Co. Alexander was in the throes of planning the design and construction of a new airplane.

Design Engineer

Barely three months later, the designation of draftsman was changed to design engineer as the result of his successful design of the Eaglerock Biplane (powered by an OX-5 engine) which became one of the more successful airplanes (Fig. 15-2) of its day. It was the airplane in which Ralph Harmon, later president of Mooney Aircraft Inc., took his first dual instruction.

Fig. 15-1. Al Mooney and sketches of the 23 aircraft he designed and helped build.

Chief Engineer

In April of the following year, 1926, Al became chief engineer at the Marshall Aircraft Co. Only 20 years old, he designed the OX-5-powered Montague Monoplane (Fig. 15-3). It first flew in September of that year. This was a high-wing fabric-covered airplane which would subsequently also be known as the M-2.

Fig. 15-2. The M-1, the Alexander Eaglerock biplane, was Al Mooney's first design.

Fig. 15-3. The Marshall Monoplane (the M-2), powered by an OX-5 engine and first flown by Joe Hammer on September 6, 1926, was Al's second design.

Project Engineer

A few months later, in December of 1926, Alexander Aircraft hired him back. He became the project engineer for the certification process of the Eaglerock and then the chief engineer. He stayed with that company until the spring of 1929 and designed a variety of advanced versions of the Eaglerock (powered with Hisso, J-4 and other engines) and the Eaglerock Bullet which, powered by an Anzani engine, first flew in June of 1928 (Fig. 15-4).

Mooney Aircraft Company

In April of 1929, the 23-year-old Mooney, now with four airplane designs to his credit, formed his own company—the Mooney Aircraft Co.—which eventually employed 20 people. Here he designed and built the M-5 (Fig. 15-5). It is a sleek low-wing monoplane which bears a slight family resemblance to the Eaglerock Bullet. Powered by a Kinner K-5 engine, it had a wing span of 36 feet 6 inches, a length of 21 feet and an empty weight of 1,215 pounds. In April of 1930, this airplane made a record flight from Burbank, California, to Fort Wayne, Indiana.

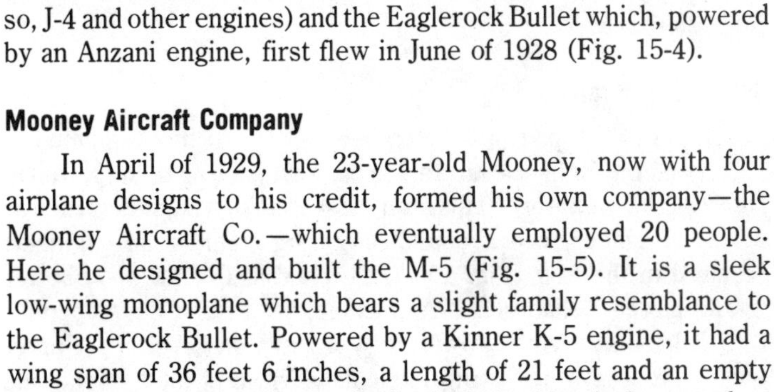

Fig. 15-4. The M-4, the Eaglerock Bullet, powered by an Anzani engine. It flew in June of 1928.

Fig. 15-5. The M-5 with its young designer, Al Mooney. Powered by a Kinner K-5 engine, it made a record flight in 1930 from Burbank, California, to Fort Wayne, Indiana.

Also in 1930, he built a long-range version of that same airplane. This one had wheel fairings and was powered by a Kinner B-5 engine. But this was the depth of the Depression, and keeping a company operating became increasingly difficult. Al and his brother moved back to Denver and under a partnership, called Mooney Aero Engineering, Al designed and they eventually built the M-6. It was an ultra-light monoplane that was powered by a Continental A-37 engine. This airplane first flew in April of 1932 (Fig. 15-6).

Bellanca Aircraft

His next stop, from December of 1931 to the beginning of 1935, was the Bellanca Aircraft Corp.—which was then located in New Castle, Delaware. This stint, first as design engineer, then as project engineer, and eventually as chief commercial engineer, resulted in the M-7 (Fig. 15-7), M-8 and M-9. The M-7, better known as the Bellanca C-27A, was a cargo aircraft developed for the U.S. Air Service. Powered by a Pratt and Whitney Hornet engine, it was a high-wing monoplane of steel-tube, metal and fabric construction. The M-8 (Fig. 15-8), nicknamed the Bellanca

Fig. 15-6. The M-6, Mooney's ultra-light monoplane. It first flew in April of 1932.

Fig. 15-7. The M-7 was a cargo aircraft designed for and built by Bellanca.

Irish Swoop, was a low-wing, single-seat monoplane in which pilot Jim Mollison set a transatlantic record. The M-9 was a somewhat altered version of the M-7 and was known as the Bellanca Cargo-Aircruiser. Powered by a Wright Cyclone engine, it was destined primarily for use in Northern Climates.

Monocoupe Corporation

In 1935, after a brief stint with a firm of engineering consultants during which he was involved in the preliminary design of a proposed 90,000-pound flying boat, Al joined the Monocoupe Corp. Here he designed the M-10 (Fig. 15-9) series of aircraft, generally known as the Dart and the Monocoupe MonoSport (Fig. 15-10), a sleek low-wing monoplane, which became a favorite with aerobatic pilots of the era. He also designed and built the M-11, known as the Monocoupe Twin Monocoach, his first twin-engine airplane which was powered by two Lambert R-266 engines and in another version by two Menasco engines and first flown by Clare Bunch in June of 1936.

Culver Aircraft Corporation

Then came seven years of a fruitful association between Al Mooney and the Culver Aircraft Corp. The result was six airplanes

Fig. 15-8. The M-8, the Irish Swoop, was also designed for and built by Bellanca and it eventually held the transatlantic record until 1943.

Fig. 15-9. Three M-10s in flight.

(M-12 through M-17). They were popularly known as the Culver Cadet (Figs. 15-11 and 15-12), the Culver PQ-8—many were used by the Army Air Force as target drones (Fig. 15-13)—a Navy version designated the Navy TDC-2 and a revised version of the

Fig. 15-10. The M-10 MonoSport.

PQ-8, designated the PQ-9, that was designed for higher missions than could be performed with the previous model. This design was later replaced by the military with the Culver PQ-14 (Fig. 15-14)—first flown early in 1943 by Earl Hartman. The last in the

Fig. 15-11. The M-12 production model.

Fig. 15-12. The M-12 Culver Cadet with pilot Foster Lane.

group was the Culver V (Fig. 15-15). It was a two-seat retractable low-wing monoplane which was first flown in September of 1945 by Earl Price (with Al Mooney in the right seat).

Mooney Aircraft Incorporated

The Mooney Aircraft Inc, was formed in 1946. First with Al, and later on without him, it survived until 1970. For all practical purposes, this was the time during which the Mooney airplane, as we now know it, saw the light of day. It started with the design and construction of the M-18—the Mooney Mite (Fig. 15-16). The Mite was the first Mooney airplane to be equipped with the famous all-moving tail assembly. This feature eliminated the need for drag-producing trim tabs.

A fast single-seat airplane with retractable gear, it was initially powered by a modified Crosley engine (Fig. 15-17) and then, in later versions, by Lycoming or Continental engines (Fig. 15-18).

Fig. 15-13. The M-13 radio-controlled drone aircraft known as PQ-8. It first flew in 1940. Large numbers of these were built for the military.

Fig. 15-14. The M-16. The military designation was PQ-14.

Fig. 15-15. The M-17 Culver V is a two-place retractable monoplane.

Fig. 15-16. The M-18 Mooney Mite powered by a modified Crosley engine.

Fig. 15-17. The Mooney modified Crosley engine.

The first flight of a Mite took place on May 17, 1947 with "Ol' Eagle" Bill Taylor at the controls. Bill Taylor test flew all the subsequent versions of Mooney airplanes for as long as Al remained with the company. A military version of the Mite, dubbed the Cub-killer (Fig. 15-19), was developed and outfitted with 30-caliber machine guns (the M-19). In 1952, Al started the preliminary design for the M-20 (Fig. 15-20). The M-20 design was the basis for all Mooneys built since, including the 201 and 231 (but

Fig. 15-18. The M-18C is another version of the Mite. This one is powered by a Continental engine. It is shown here with test pilot Bill Taylor at the controls.

Fig. 15-19. The M-19, a slightly larger version of the Mite, was designed specifically for the military and was known as the Cub Killer.

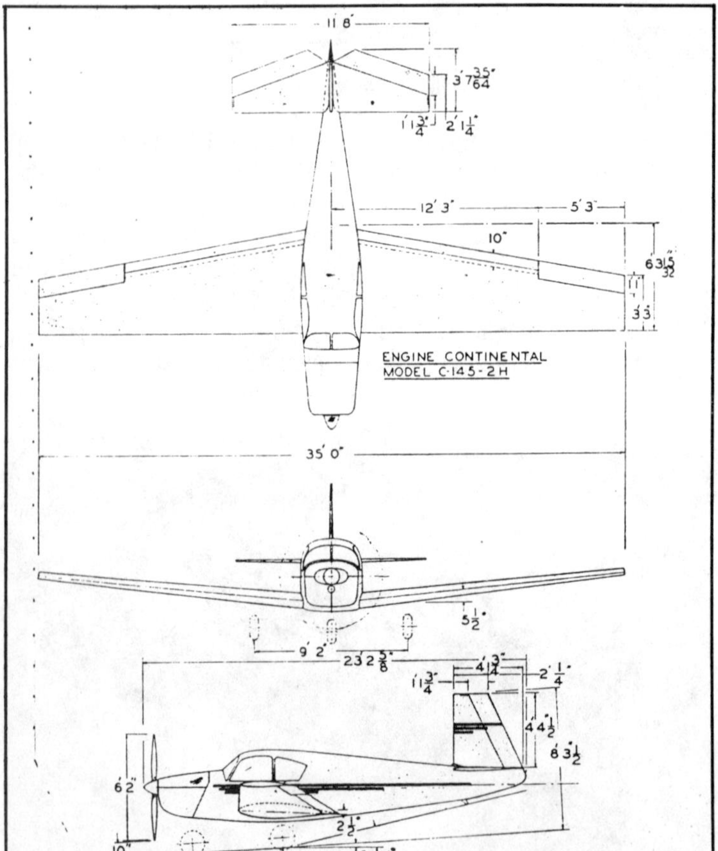

Fig. 15-20. In 1952, Al started preliminary design on the M-20.

Fig. 15-21. The first prototype of the M-20. This model was later known as the Mark 20.

not including the Mark 22 Mustang). The prototype of the M-20 (Fig. 15-21), powered by a Continental C-145 (145 hp) engine, was first test flown by Bill Taylor in September of 1953. It was a wood-wing airplane with an all-moving tail section and a manually operated retractable gear. Al Mooney said that even then he envisioned replacing that with an electrically operated gear. From the very beginning, it was known for speed coupled with exceptional economy (Fig. 15-22).

Once Al had obtained type certification for the M-20, the people who had financed the company and owned all the stock, according to Al, apparently thought that he had outlived his usefulness to the company—resulting in his leaving. Another version of the story claims that he got an offer from Lockheed which Mooney management was not financially able to match. Be that as it may, in 1955, Al Mooney joined the Lockheed-Georgia Company in Marietta, Georgia where he designed and supervised production of

Fig. 15-22. The production version of the Mark 20.

Fig. 15-23. The M-21, the Lockheed LASA-Sixty aircraft, later manufactured under license in Italy, Argentina and Mexico.

the last three airplanes to carry the famous M designation—the M-21, M-22nd M-23. The M-21 was a single-engine, high-wing utility aircraft that looked pretty much like a Cessna. Dubbed the LASA Sixty (Fig. 15-23), and powered by a Continental IO-470 engine, it first flew in September of 1959. It has subsequently been built under a production agreement in Italy, Mexico and Argentina.

This was followed by the development of two versions of a military aircraft. The M-22 was known as the Lockheed Hummingbird I (Fig. 15-24). It was a VTOL research aircraft with ejector lift augmentation and was powered by two Pratt and Whitney JT 12A-3 engines. The aircraft made its first conventional flight on July 7, 1962 and its first hover flight in September of that same year.

The M-23, better known as the Lockheed Hummingbird II (Fig. 15-25), was a more powerful 6-engine version of its pre-

Fig. 15-24. The Hummingbird I in flight.

Fig. 15-25. The Hummingbird II on the ground.

decessor. It first flew on September 28, 1968. By that time Al Mooney (Fig. 15-26), now 62 years old, had retired from the company.

It had been 43 years from his first day as a draftsman for the Alexander Aircraft Co. in Denver to his last day at Lockheed-Georgia in Marietta. The result was 23 aircraft designs—or one every 1.9 years—not bad.

Semi-Retirement

Actually, his retirement, at least for a time, was really only a semi-retirement. Known far and wide for his expertise in aeronautical matters, he was sought after as a consultant. He intermittently performed these services for the Wren Aircraft Corp. in 1968, for Burns Aircraft Corp. in 1969, for Continental Motors and AiResearch Manufacturing Co. in 1969, and for VIM Corporation and for Lockheed-Georgia in 1970. In the years since then, he was elected to the Aviation Hall of Fame, the OX-5 Aviation Pioneers Hall of Fame and the Colorado Aviation Hall of Fame.

The Old Days

During a visit to Kerrville in the process of assembling background material for this book, I spent several hours with Al Mooney and Art Mooney in their home in Ingram. They talked at length about the old days.

Al: "It was during the war and we were at Culver building Cadets when one day some general, I can't remember his name, walked in and said, 'How about if you take that little Cadet there and put a tricycle gear and a big generator on it?' I said, 'Yeah, why?' 'We want to buy a bunch of them and use them for drones.' 'Sure, we can do that.'

"Well, I went to Continental and they weren't interested in putting a big generator on their engine, so I went to Lycoming and they weren't interested either. Then finally went to Franklin and they said, 'Sure we'll do it.'

"So we bought a bigger building and I was doing my darndest to get the first drones built, but old man Culver wasn't too interested in the drones... He liked building commercial airplanes and couldn't care less about the military. And then, one day that general called again, Harris, I think, his name was, asking when we're going to get some of those drones built. 'I'm doing my darndest,' I told him, 'to get those guys around here to get on this, but all they want to do is build commercial airplanes.'

"So, suddenly a lot of heat came from somewhere. Here we had a nice, big military contract, a place in which to build the airplanes and a good crew of guys, but the people upstairs couldn't get interested. Then one day some brass showed up just before Christmas and said, 'If you can get five airplanes out between Christmas and New Year, that's what we want.' So we got the boys together and worked around the clock and actually were able to deliver five airplanes by New Year. 'Oh, so you can do it after all,' they said to us. 'How many are you going to deliver next week?' We told him he'd be lucky if he got any next week.

Fig. 15-26. Al Mooney (right) and Art Mooney with the M-23 under construction.

"Anyway, over the years we built thousands of those drones. PQs they called them. They could be flown by a pilot or be operated by remote control, and that was good because you could ferry them to anyplace where they were needed.

"Meanwhile, I had been fooling around with an advanced design, but I wasn't necessarily thinking of actually building the airplane. The idea I had was for this little single-place airplane. I thought that all the guys who had been flying during the war—when they came back, they'd want to keep on flying—so this thing, being cheap, ought to sell pretty good. And later on it did, too. Altogether we built and sold about 275 of the Mites.

"But I'm getting ahead of myself. I remember, on V-J Day I was upstairs in the office getting all kinds of wires cancelling the military contracts while everybody else was celebrating. And then, for a while, things didn't look too good—except one day a guy I'd never really heard of showed up and introduced himself as Mr. C.D. Yankey, Charley Yankey. 'What's on your mind,' I asked him. 'Well, old Walter (Beech) and I have decided to operate this place for you.' 'You mean, take it over?' 'Yeah.' 'And you already agreed to this?' 'Year.' 'Well, you must be a dumb jerk. He flared up, 'What do you mean?' 'You should have looked into this place and found out how much it owes before you made that agreement.' He got made and stalked out.

"Next day Walter came over. 'Al, what is the matter with you?' And then I found out that they had already paid every bill, so I told him, 'Gentlemen, I'm your boy. I'm sorry I got nasty yesterday.' And the old boy said 'I don't blame you.' And that was my start with Pappy Yankey.

"Anyway, we set up this company called Mooney Aircraft in Wichita and started to build Mites and also a military version of the Mite, the M19, with which you could fly along the ground, jump over trees and be gone. They called it the Cub Killer. Anyway, we had this little factory not far from the airport east of Wichita, and about 1953 there was that Korean emergency, and the airport was being taken over by the military and we were going to be stopped from flying off that field, and we knew we'd have to move someplace. That's when George Haddaway steered us down here to Kerrville."

Art: "He raked up $25,000 from small businessmen in Kerrville. The big ones, like the Schreiners, they didn't want any manufacturing down here. They had nice cheap wages and they wanted to keep them that way. Another problem we'd been having up in Wichita was, every time we got a guy trained so he was a pretty good workman, Cessna or Beech would hire him away and pay him more than we could."

Al: "Anyway, George fixed it up for us and we moved down here and started building Mites again. It was around that time that old Pappy (Yankey) began to lose his health.

"Meanwhile I had got the idea for a little 4-place, an advanced aerodynamic design. We had a heck of a time getting that airplane built (the M-20), but when it was built it was way ahead of any other airplanes in its class in speed and fuel consumption. Old Pappy came down here in his big Chrysler station wagon and when he saw the airplane he got all charged up.

"But that was the last time we saw him. Shortly after he just faded away and died. He left me all his stock—he had half and Art and I each had a quarter—with the single provision that I had to get the company financed. You just try that—especially back then. Eventually I got it financed all right, but we had to give away all our stock. That was when Hoffman and Rachel came in."

Art: "They just suddenly showed up one day on their own."

AL: "Actually, Pug Piper had also come down to see us during that period. That's when we blew it. He was ready to give us a straight contract to build 100 Mites a year, but we would have had to agree not to build another 4-place. I didn't want to do that, so I finally made the deal with those guys. I agreed to stay for the duration, to get the airplane certificated, and when that was done I guess they felt they didn't need us anymore. So I went to Lockhead with the understanding that I could assemble my own crew. That was the beginning of what we called the Possum Works at Lockheed, and I stayed there until I eventually retired."

Chapter 16

Ralph Harmon

Ralph Harmon (Fig. 16-1) was born in 1913 and he traces his career in aviation back to his teens.

"I guess my career started with building a Pietenpol that was published in *Mechanics Illustrated* in 1930 or so, and I learned to fly in it. I soloed in it in 1933 and I've been flying ever since. Actually, I did have some dual, two hours and 50 minutes of it, in an Eaglerock (the Eaglerock was Al Mooney's first design, the M-1.), plus three hours and some minutes of solo.

There wasn't anything to getting a pilot's license in those days. You simply wrote in for an identification number, I still have the certificate, number 12866, and then you could go fly the mail, go barnstorming, do anything. I used to go barnstorming in my airplane, fly it in air shows, did all kinds of things. Then I started to work around various small companies in Indiana and Illinois. During that time, I'd designed a 2-cylinder opposed engine with the camshaft over the top of the engine. The idea was later picked up by Continental in the Tiara engine. And one day I sent the drawings to Beech Aircraft to apply for a job.

"So they hired me. The job paid 60 cents an hour and I moved myself and my family from Indiana to Wichita and started at Beech as a draftsman. I stayed at Beech for 14 years, became a layout designer, then got into the advanced design group and was put in charge of preliminary design, then became chief design engineer and by the time I left I was assistant chief engineer.

The Beech Bonanza

"During that time, I wrote the original specs for the first Bonanza, made the preliminary sketches and presented it to the management and they put me in charge of the Bonanza program. Eventually, I was also in charge of the Twin Bonanza and T-34 programs—plus several other airplanes that we built experimentally, but that never got into production.

The 620 Program

"And then, in '53, I left Beech and went to Cessna in charge of the 620 program. We set up a whole new experimental division to develop this transport category pressurized airplane. We actually built one and had it almost certified and Cessna started building the Wallace Plant to build that aircraft. But just then the government dumped Convairs and DC-3s on the market for $20,000 and $30,000 and our airplane was going to cost $350,000, and we just felt that we couldn't compete with them. So that project was discontinued.

"While I was at Cessna I was also a consultant for McDonnell Aircraft. McDonnell was in the military business and Cessna was commercial, so the two companies exchanged consulting services, because Cessna wanted to get into the military business and Mac wanted to get into the commercial business. So I was the commercial exchange consultant on the McDonnell 119. When the 620 program was cancelled, Jim McDonnell contacted me and asked me to come to work for them. That was in '57. So I did.

"But, I'm afraid, we didn't have a winner there. The JetStar was a much better airplane. We spent some time trying to convince Mac that he didn't have a winner. At that time, I had about 550 engineers in my operation. Eventually, Mac decided that he wanted to get into the airline business so we started design studies on something close to what the DC-9 is now. But I didn't feel that I wanted to get into that business. I'm a small-airplane guy. If I can't fly it I don't want it.

"In the meantime, when I had first come to McDonnell, George Haddaway had sent Hal Rachel to me with an offer to join Mooney down in Texas. Well, I had just gone to McDonnell and I wasn't going to change. Then, after I'd been there a couple of years and saw how things were going, George got back in touch with me, and then Rachel contacted me again, and so I came to Mooney.

The M-20A

"I came down to Kerrville in January of 1960. At that time, as I recall, there were about 170 employees and the engineering department consisted of, I think, three people—none of whom had

Fig. 16-1. Ralph Harmon, vice president of engineering and later president of Mooney Aircraft Inc.

ever worked in an airplane factory before. Some of them were aeronautical engineers, but none had practical experience. Actually, there wasn't anybody in the factory who had ever worked in an airplane factory before. It wasn't easy, but we started. They were making the wooden-wing airplane at the time, the M-20A, and there were a lot of problems with it. Wood is not a good material for production. It's easy to build a prototype a low cost, cut to produce it afterwards presents a lot of problems.

"I could see a great potential in the airplane if it was modernized. So I started right away, in January, to teach these guys what the heck an extrusion was, and an A-D rivet, and 20-24 aluminum, and I mean from zero, almost.

I personally hadn't done a stress analysis or detail design for years and I went right back to the board and hit the books. I also had to pretty much design the jigs and buy the equipment. But we did it. We had a heck of a group and everybody got with it. We got a type certificate for the all-metal M-20B on, I believe, the 21st of December of that same year.

"We had a show at the auditorium, had the airplane up on the pedestal, and we had the people in the audience at the show to watch the demonstration. My chief engineer came running down the aisle with the type certificate, and, boy, we had to have it, too.

"Then we started building them. First the B—which was a straight conversion from wood to metal. It was kind of heavy and there were a lot of things that weren't too good. So then we went to

Fig. 16-2. The Executive was a stretched 200-hp version. It is basically the same airplane that is still being manufactured today.

the M-20C. By that time, we had time to kind of refine the thing and take a second look at it. That's still being built today. That's the Ranger. I've still got one. Then came the Super-21, which had a 200-hp engine, and then the Statesman which didn't sell too well at the time but is very popular today, and then the Executive which is the 200-hp version of the Statesman (Fig. 16-2).

The M-22

"Then we got into the M-22 program, the pressurized single, which later was called the Mustang. Actually, back in 1950, just a day or two before Walter Beech died, he came to my office and I was talking to him about building a pressurized single-engine airplane. But when he died, that was the end of that. So I'd wanted to build one all along, and at this point Hal Rachel was very interested in the idea and so was the sales department because at that time nobody had built a pressurized light general aviation airplane. Anyway, Hal was intrigued and he said, 'You go ahead and build it, and I'll find the money. Don't worry about it.'

"So we started on the program, and we built it. It was a kind of compromise, it wasn't a very pretty airplane (Fig. 16-3), but it's still a better pressurized airplane than anything comparable flying these days. In fact, its performance was almost equivalent to the Beech Duke in speed and so forth.

"We built 27 of them, as I recall, and to this day, to my knowledge, there aren't any AD notes out against that airplane.

"Admittedly, the development costs had been quite high. Not necessarily so for the kind of project it was—it would have cost

Beech or Cessna quite a bit more—I don't remember the exact figure, but we had it well tooled.

"Then Rachel made a press statement about how we were going to sell these and dominate the industry and put Beech out of business and Cessna out of business and stuff like that. I always had a horror of every news conference he ever called. I never knew what he was going to say. He told them that we'd be in the jet business and in the turboprop business, and, of course, we did get into the MU-2 program. At one time, he wanted to buy the Corvair automobile project and move it to Kerrville. I told him, I said, 'Have you ever been in an automobile factory?' 'Never seen one.' I said, 'I'm going to haul you up to Detroit and show you what one looks like.' But he wouldn't go. 'I don't have to look at it.'

Money Problems

"By this time, all kinds of money problems had developed. Financing became difficult. Still, we were in great shape to go public, and Kidder Peabody wanted to work out a deal for us to go public. But Hal wouldn't have any part of it.

"I remember, I hauled him up to New York where Dick Young, one of the members of the board of directors, had set up a deal with Kidder Peabody and another underwriting company to work out a program to go public and get some financing and pay off that...Mustang program and get off this Mooney bond business (see the section on bonds at the end of this chapter).

We had the meeting and I never will forget it. We walk out on Wall Street, Dick, Hal and I, and Hal turned around to face Dick and me and said, 'We'll never do business with those people,' or words like that. Of course, he and Hoffman owned 85 percent of the company and I had 5 percent. So there I was.

Fig. 16-3. At one point, a T-tail version of the Mustang was built as an experiment (courtesy of Ralph Harmon).

"But we got into the MU-2 program, and I've got to give Hal credit. He's a tough negotiator. He negotiated a good deal with them and we set it up, ran it, and eventually sold a lot of airplanes. Hal was smart, sharp as a tack, a genius in a way. The guy had a lot of good features. He was aggressive, inventive, clever, he was a dynamo. He ran seven days a week, 30 hours a day, like Jim McDonnell, the same kind of drive.

"The MU-2 program stayed with Mooney until the bankruptcy and then we were able to separate that program from the bankruptcy. The Mitsubishi people bought the factory and the facilities we had put up in Sand Angelo. They were nice people, great people to do business with. But for Mooney this was sort of the beginning of the end."

Bonds

That "Mooney bond business" that Ralph Harmon referred to is kind of interesting. This is how it worked:

Any time someone bought a Mooney airplane, they were given a $1,000 bond that was cashable if and when that airplane, for whatever reason, was removed from the FAA's register of aircraft. When that happened, either because of an accident or for any other reason, the holder of the bond could turn it in to Mooney and receive the $1,000. It was kind of like a $1,000 Green Stamp.

The result of this, as far as Mooney Aircraft Inc. was concerned, was that the company was carrying $1,000 per airplane of obligations on its books. That sum was non-taxable. Now, assuming that some 3,000 or 4,000 Mooney aircraft were flying around, that meant that some $3 million to $4 million dollars worth of capital was carried as a liability and was therefore excluded from taxation.

Chapter 17

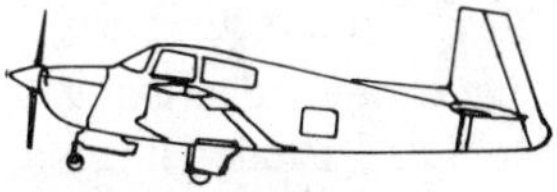

Roy Lopresti

Who is this man who, by doing what all the experts said couldn't be done, made a good airplane better?

"I always wanted to design airplanes. Like everybody else, I built model airplanes. And in the last year of school, all the students competed in designing their own airplanes. We were all competing for the Chance Vought Design Award and I won it in 1950. I decided then that I wanted to work for Piper or Cessna, but at that time there were no jobs to be had, at least, no jobs in the small-airplane business. I was really disappointed.

I finally got a job working for Grumman Aircraft and got involved in something that was as far from small airplanes as you can imagine. I worked on supersonic cruise missile—submarine launched. But luckily for me, my boss on the missile program was Dave Thurston. Dave was building an amphibian and I had a chance to help a little. I didn't do much. Dave did all the design work and all I did were some drawings. But at least it involved a little airplane. (That airplane later became the Lake.)

"After about a year of that, the Air Force pulled me back for active duty. I had an ROTC commission and decided that as long as I was in the Air Force, I'd want to fly. I already had my private license, but I went through military flight training and I flew in the Korean War.

"When that was all over, I got married and decided I wanted to settle down some place. So I want back to Grumman. But, just like before, I again got diverted from airplanes—got involved in the space program instead. I helped to design the Lunar Module and was one of the people who put together that whole proposal for Grumman. They then sent me to Cape Kennedy to set up the operation there and I became chief engineer for the Grumman operation at Cape Kennedy. It was really a fantastic job. Everything was new and exciting. Nobody really knew if we'd ever get to the moon, but we sure gave it our best shot. But the job was also frustrating.

"I had a nice house with a boat in the backyard and a Bonanza on the airport just 2 miles away. Didn't have much money, but I'd saved some. One day I got up and said, 'You know, Peg, people have been designing airplanes for years, but cockpits are just awful. I think I'll design a better cockpit.' So I went ahead and bought some plywood and started to make a cockpit mockup in the garage. I was really having some fun.

So I said to myself, 'What am I doing in the space business? I really want to be in the airplane business. Well, if I can design a good enough airplane, maybe I can start my own company.' Very naive! But it was kind of a fun idea.

An Exciting Amphibian

I did sort of a home-made market survey and arrived at the decision that there was a real need for an exciting amphibian. The only one that was being built at the time was Dave Thurston's Lake. It was a nice airplane but, like practically all amphibians, it had lots of shortcomings. So I designed an amphibian that would be just about as fast as a land plane. I raised some money and went ahead and built the airplane.

"I figured that if you can build an airplane that's obviously very good, you ought to be able to find some money to produce it. I tried for three years. All kinds of aviation people came to see it, but to make a long story short, just as I came close to making a deal with Grumman—which had some idea of building the airplane in pieces in Singapore and then shipping the pieces to the U.S. for assembly—the Vietnam War broke out, and everything fell apart.

"I never did raise the $3 million I thought I needed. Maybe the timing was wrong. Anyway, I couldn't go back to the Grumman job at the Cape, because that operation was in the process of being shut down, and I decided I'd not got back to New York (Grumman's Long Island facility) to work on big airplanes. I wanted to work for Piper or Beech or Cessna.

"Then, just about that time, Grumman Aerospace bought American Aviation. I kind of raised my hand and said, 'Hey, you guys, I've done a lot of things in the airplane business I'd like to work for you.' So they said, 'Come up to see us in Cleveland.' I went up there and they offered me a job as chief engineer, a lot more than I had expected.

The Trainer and The Traveler

"I walked into Grumman American and the job was made exactly to my liking. It couldn't have been any better. At the time, they were building the Grumman Trainer and the Grumman Traveler. They both weren't bad. The Traveler, to me at least, looked a little strange. The guys in the shop had actually put together a higher-horsepower version of the Traveler around a 235-hp engine. The reason they did that was because the marketing people there thought that anything that Cessna did was absolutely right. And if Cessna used that engine, it had to be pretty good. So they put together this higher-horsepower version of the Traveler which they called *Fat Albert*. It had a big, ugly cowling and some tapered sections over the in-board portions of the rectangular wings to accommodate retractable gears at some later time.

"Anyway, I said, 'Let me fly your new airplane.' I did most of the flying in those days. That was Monday morning. On Wednesday we had our first staff meeting. Everybody was there, Russ Meyer, the president of the company and a lot of other VIPs. So Russ Meyer says, 'What do you think of our new airplane? Remember, I'd been there just two days. But I had to tell them: 'It'll be very difficult to bond that tapered wing, and the airplane doesn't fly very fast, and it's ugly as sin. I think you ought to throw it away.'

"At first, they were kind of ticked off at that, but the final outcome was, 'If you're so smart, go ahead and design us a better one.' And that's how the Tiger started.

"And, I guess, that's my life's story prior to coming to Kerrville."

Chapter 18

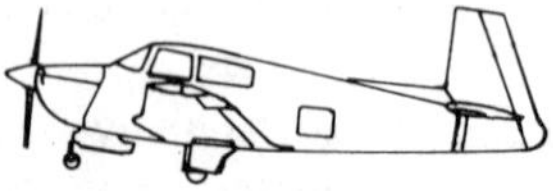

George Haddaway

One of the questions I kept asking when I first started to research this book was: "Where is Al Mooney?" But no one seemed to know. There were all kinds of answers like "I think he moved down to Florida," or "The last I heard he was working with Burns on that twin somewhere in Mississippi or Alabama," but nothing definite. I knew that if this book was to be complete, I'd have to find him and talk to him because, after all, it was he who had created the original design (the M-20) on which all subsequent Mooneys were based.

And then, purely be accident, I ran into George Haddaway at an aviation convention in Las Vegas. George had been the editor and publisher of *FLIGHT* magazine for some 40 years and I remembered having heard that he had known Al Mooney in the old days. Sure enough, when I mentioned my problem, he not only knew that Al was retired in Ingram, Texas, a small town just a few miles from Kerrville, but he even had his phone number.

"You know, of course, that it was I who arranged for him to come down there to Kerrville when that airport they were located at in Wichita was being turned over to the Air Force."

I hadn't known, but right then and there we decided to get together and let him tell me about it.

"In those days, I guess it was somewhere around 1952, Pappy Yankey was still alive (Charles G. Yankey, a wealthy Wichita attorney who, at one time or another, provided backing for Walter Beech and for Clyde Cessna and who, at that time, owned 50 percent of the stock of Mooney Aircraft Inc.) and with Mooney having to get out of its facility in Wichita, he asked me to find them a home. One of the first people I contracted was Jim Wright who was then mayor of the town of Weatherford, some 20 miles west of Fort Worth. (Jim Wright is now a Congressman from Texas and Majority Leader of the House.)

The reason I took this deal to Weatherford was because we had the Convair Division there and there were a lot of people out of work at the time—aviation people—and if Weatherford would take this on, there would be all those people available. So I told Jim that if he'd arrange for us to have a spot on the airport and get some financial backing from the townspeople, he'd have himself an aircraft factory there.

Well, Jim did want that aircraft factory—he was a pilot in World War II—and he asked that I show him the financial data, and I did. Well, frankly they didn't look too good, and he came back and said, 'George, I took that stuff to the bank and they slammed the door right in face. There's no way you can come to Weatherford with Mooney.'

The Search

"I then took it all over the place, Magnolia, Arkansas, all kinds of places, and Kerrville was one of them. There the local people showed a lot of interest and the main attraction was the deal we got on the airport—a lease for a dollar a year for 99 years. Well, Pappy Yankey bought that right off, grabbed the bait and hid under a rock. And if Pappy hadn't died, there would never have been all that mess with the company that developed later."

But Charles Yankey did die and he left his 50 percent of the stock in Mooney Aircraft Inc. to Al Mooney. His heirs decided that they wanted nothing to do with the aviation business and that effectively left Mooney without a source for the necessary funds to complete the certification process on the M-20. Only one M-20 had been built so far.

A Last Minute Plane

At this point, Haddaway called Al Mooney to find out how things stood.

"I said, 'Al, what're we going to do?' and he said, 'I have to declare bankruptcy.' I said, 'Over my dead body.' So I got into the little Cessna 170 and I flew down to Kerrville. It was Saturday and they were prepared to file bankruptcy on Monday morning. They owed about $350,000, which Pappy would have paid if he'd been alive. Of course, they had nothing coming in because the single-seater, the Mooney Mite, was out of production and all the chips were on the table for the M-20.

"I said, 'Al, you've never been bankrupt.' I said, 'You can't afford to have that on your sheet.' Well, he says, 'I have no other alternative. The Yankey estate wants nothing to do with airplanes.' So I went to the lawyer's office upstairs over the Schreiner Bank. He had the bankruptcy papers. I said, 'Let me look at them.' So I took a look at them, and they were supposed to go the Austin on Monday. So I said, 'I'm going down stairs and copy some of this.'

I went downstairs, went out the door, took a car or a cab or something and went out to the airport, jumped in my airplane and took off and spent the night in Midland, Texas. I had a rich cousin out there, and I figures maybe I'll get Al some of that oil money. I landed at Norm Hoffman's airport. He and I went out to eat together and he brought his brother-in-law, Hal Rachel. So I told them, 'Boys, we have a bird's nest on the ground down there in Kerrville." And, yes, they knew all about it, I said, 'You can steal it.' They said, 'We heard they were going bankrupt.' I said, 'Here are the bankruptcy paper. They can't go bankrupt without them.'

"So they went down there and they flew the prototype M-20. Hal Rachel fell in love with the airplane. He asked me, 'Are you sure we can get this real cheap?' I said, 'Yeah.' 'But,' he said, 'there are $350,000 out against it.' I said, 'Yeah, but most of that you can settle for 10 cents on the dollar. I know most of those people and they'll go along because they all want to see Mooney succeed.'

The Deal

"Anyway, to make a long story short, they made their deal with Al. Actually, Hal and Norm had a fueling contract with the Air Force and they were in a very high tax bracket. Hal, being a tax lawyer, figured that they were actually getting into this deal for nothing, which they did.

"And I have to give them that, they really did a good job. They got into production, they built up the factory, got all the tooling. Art Mooney, Al's brother, was out in the shop and Al was on the drawing board. There are so few people in this business, beginning with the Wright Brothers, who can design an airplane, then build it with their own hands, and then fly it. This was the tradition that those two people represented, And this, to me, was the important thing.

"So, with everything going, I phased out of the deal. But I did continue to go to their sales meetings. At the last one I attended, I could see the handwriting on the wall. Rachel was running the company. Hoffman was absolutely in another world. He was out selling and peddling and building up the distribution. But something was beginning to go wrong. It was a bitter disappointment for me. Here these guys had the world by the tail with a downhill pull, but somewhere along the line they got greedy.

"You know, of course, that right after Al had completed the certification of the M-20 they let him go—which was awful. Granted, Lockheed then hired him and Art, so they were all right, but it was still a lousy deal.

Chapter 19

Norm Hoffman

Norm Hoffman, with his brother-in-law Hal Rachel, took possession of Mooney Aircraft Inc. The two of them ran the company for the next 14 years.

A Good Deal

"One day Hal and I were having breakfast down at a hotel in Midland, Texas. We had been in the aircraft business ever since 1947. We were just sitting there having breakfast and George Haddaway came in. He started talking to us, visiting with us a little bit, and George said, 'You know, you fellows ought to take over that Mooney Aircraft. It needs somebody that'll push it, somebody who's got the go, who can get it going again. It's a good deal.' At first we didn't think too much about it, but then we did agree to go down and look at it.

"By that time, Al Mooney had the 4-place prototype, I'd say, pretty close to finished, or it may even have been flying, I don't remember. But it was just a rough airplane. And everything was shut down. Nobody was working. There just were Al and Art and just a few of the people, just kind of hanging on. And we got down there, Hal and I, and we looked it over pretty good. We went back to Midland and somewhere along there we sat down and discussed it some more.

I remember, Hal made the remark, 'Well, it will not be easy. It's going to take a tremendous amount of effort and time from both of us. And the possibilities of ever making it go are not really that great. So, what do you want to do, Norm?' And, of course, being young and energetic and not knowing really what was going on, I said, Let's do it, let's do it." All I knew was that it was a chance to get into a part of aviation that I hadn't had yet. So we shook hands and said, 'Okay, let's go in,' and we jumped in with both feet without really knowing exactly what we were jumping into.

"We had been fairly successful in running a couple of fixed-base operations in Midland. About two years prior to this, we had gotten a contract for the servicing of military jets amounting to 12 million gallons of fuel a year. We'd made some money on that. So, having that running, plus the two operations in Midland, we thought we just didn't have quite enough to do.

"So, anyway, we decided that if we could buy it the way we wanted to buy it, we could take the money we'd made in Midland and from the gas contract—plus we had an opportunity to sell the gas contract—we could take that money and handle the necessary financing to get it started. And we just jumped in with both feet and started to work.

The First M-20

"Al worked on getting the M-20 certificated, which took a little longer than he had thought, but then that's standard operational procedure in this business. In the meantime, we had all kinds of bits and pieces for the Mite. So we put together some Mites, got the factory started, started training people, and slowly started building an organization.

"The first airplane that came out, the first M-20 after we got certification—I took that number one airplane while they were getting engineering and everything organized; I took that airplane and I started on a tour, flying around the country, trying to set up distributors and make presentations with the aircraft to people who were distributors for other airplanes whom we felt we wanted as distributors.

"In the meantime, as I recall, Lockheed had a program for which they wanted Al. They made him an offer. I can't really recall how far down the road we had gotten with the airplane, but it wasn't very far. Anyway, we couldn't compete with Lockheed's offer, so Al went to Lockheed. It wasn't too bad. He really had done what he was supposed to, and he had built a skeleton engineering crew, and so we just sort of stumbled along.

"Eventually we got ourselves pretty well organized—started to build a little better airplane, something that would sell, getting some of the bugs out of it. And then we started looking for an engineer.

An All-Metal Airplane

"Ralph Harmon, at that time, was working for McDonnell in St. Louis and we started negotiating somewhere around 1958 or '59. He came on board and the first thing we started to do after Ralph got there was to make an all-metal airplane. And that conversion was completed by, I think, 1960. Like with any new airplane, there were a lot of things wrong with it. People would tell us what was wrong with it, but we had a very good performing airplane for the size, so we sold them. As best as I can remember, during the time I was with the company, we sold about 5,000 airplanes all over the world (Fig. 19-1).

"During this time, we had developed a small but very enthusiastic distributor-dealer organization and we wanted to come up with another airplane for our distributor organization. At first, before Ralph Harmon joined us, we thought about building a little twin. We had even put one together and showed it at one of our distributor meetings. Maybe that's the airplane we should have built.

A Pressurized Single

"But Ralph had his own ideas and he was a strong character. He convinced us that we should build a pressurized single-engine airplane rather than the twin. I guess that was one of our mistakes. It eventually helped to tear us down. Either Ralph didn't have the capability, or we didn't have the money and the capability. And on top of everything else, there was a recession in aviation—so the timing was bad. All in all, this program, which was supposed to have cost about $1 million, ended up costing something like $4 million.

"And then, at that same time, the Japanese came after us and wanted someone in the States to market the airplane they were developing.

"I took that job over. I went to Japan and looked at the airplane—flew in the airplane. I told them what I didn't like about it, then came back and told Hal and Ralph what I'd seen and what I didn't like about it. And the only way that I could see that we could accept it and take it under our wing and sell it as a Mooney airplane, which was if certain changes would be made.

We had strong enough image at the time to be able to force that. So I went back and negotiated a contract. I had them pull the engines which were on the airplane, and the propellers, both of which were French, and had them replace them with American

Fig. 19-1. Over 5,000 of the all-metal Mooneys were sold (courtesy of Mooney).

engines and propellers and perform the necessary redesign work. All of that took time and money.

"So, now we were developing a pressurized airplane and we had a twin turboprop and we really had our hands full. We had two real bears by the tail. The Mustang, as we decided to call the pressurized single, somehow looked like something that had been modified from something else. It simply didn't have the beauty you look for in an airplane. And then we had the MU-2 which was anything but salable.

I had asked for a quiet airplane and it was noisy as anything. We had the Japanese image to overcome. And I remember the first two airplanes they brought over here. They delivered to AiResearch in California to have interiors installed. When I saw those interiors they had ordered, I had 'em strip them out. I had them redo all the interiors before I would show the airplanes for the first time.

The First Showing

"I'll never forget that first showing in San Francisco. We had a press conference and the next morning we had people coming out to the airport where we had the two airplanes on display. So, when I came out there, there were the two airplanes plus about 25 white-coated Japanese scurrying about. I found the manager who was running the show for the Japanese and I said, "What are those people doing here?" He said, 'Oh, they have to keep the airplanes going.' And I remember saying to him, 'Well, if it takes that many people to get these things to fly, we might as well get rid of the airplanes right now, because nobody is going to buy 'em, or you get rid of those people and let's sell the airplanes.'

He was very upset with me. But I told him to get rid of those people, that I didn't want to see even one of them anywhere near the airplanes, 'Otherwise, let's just send the airplanes back to Japan,' or something to that effect.

"We demonstrated the airplane that day and it was so...noisy, absolutely impossible, that that night we went down to San Jose to one of my distributors and we spent all night long making double windows for those airplanes. We put double windows in before we ever let anyone else get into them. And, of course, they were very difficult to fly. There we had an airplane that wouldn't sell, we had a Mustang which we had promised to sell for $29,000 and which was costing so much to produce that we'd have to get $50,000 or $60,000 for it to come out.

Somehow Harmon had missed the boat on everything that you can come up with on that airplane. Money was flowing out the door faster than we could borrow it. Our reserves were going down and down. We built factory in Sand Angelo for the MU-2, built additions to the factory in Kerrville, and somewhere along the line our dealers convinced us that we needed a training airplane so we bought Ercoupe—traded stock or something. Basically, I guess you might say that with all that attendent pulicity, we were just trying to hide our problems.

"Meanwhile the MU-2 still wouldn't sell. It took over two years from the time we first showed it to the time we brought out the F model which finally proved to be salable.

Bankruptcy

"One day our bankers and Ralph Harmon got together and convinced me and convinced Hal that he should resign. That would make the bankers happy and that, if he would do that, they would see to it that the company would not go into bankruptcy. Well, we agreed to do that. And we had no more than released the company and made Ralph Harmon the president when we were in bankruptcy.

Hal was indicted. I don't really know why, but anyway that's what happened. And then, in bankruptcy court down in San Antonio the company was sold to some company from Pennsylvania. (American Electronics Laboratories). So when that company took it over, Ralph was running the show and I was still in San Angelo. We still had that contract with Mitsubishi. And I remember, one day Ralph came into the office there where I was to tell me that it was my time now. I said, 'Oh, okay, there's the chopping block, is that it?' He said, 'Yeah, I'm afraid you have to go.' And I just left . I figured I'd had enough.

"As I look back on it now, I look at it as a part of my life I gave to aviation. I think that we developed a very fine airplane. It cost me 14 years of my life for nothing. Not one thing did I get out of it. I don't know very many people who have worked 14 years to get nothing. Because during that 14-year period I never received a nickel in remuneration. Both Hal and I never received a nickel. Well, I take that back, I did sell a little stock that I had owned, but neither Hal nor I ever drew any salary from Mooney Aircraft throughout those 14 years.

Very few people will believe that, but it's the truth. Of course, the money that we had invested we lost. I don't remember how much that was, but it was quite a lot. Hall, of course, lost more than I because he'd put more in originally.

"I guess, if you want to analyze it, we just went into it with not enough money. There just never seems to be enough money. And not being a public company, it was always difficult to get additional capital.

Oh, I suppose there were times when we could have gone public, but whenever we had the opportunity, things were going fine and we didn't feel that we needed to go public. Then, when we wanted to do it, things just weren't going well enough for someone to want to underwrite a stock issue. Granted, there were times when we could have gotten out from under. People wanted to buy the company and we didn't have enough sense to sell it, or if we had enough sense to sell it, we were trying to get too much for it.

"I guess the only thing I did get out of those 14 years was a...good education, an education which has made it possible for me to make a living in the aviation business ever since."

Today Norm Hoffman is in the business of selling all kinds of large airplanes through a company called Interjet. It is located in an elegant hangar and office complex at Addison Airport some miles north of Dallas.

Chapter 20

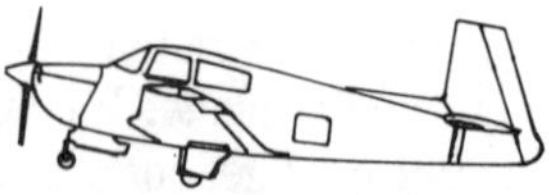

Jack Womack

Mooney's advertising, from the very beginning, was handled by an Odessa, Texas, advertising agency called Womack/Claypoole—and more specifically by Jack Womack. As a result, Jack was close to the company for the entire 14 years from its inception until its demise in 1970.

"I started in 1956, soon after Norm and Hal had taken over the company. At that time, they were still operating West Texas Flying Service in Midland and that's how we made the connection. He called me to do a little brochure for West Texas Flying Service and then he asked, 'You want to handle the Mooney advertising?' I told him, 'Sure,' and he said that they had a $10,000 budget for the first year, which wasn't much, but at least it got us started.

The First Ad

"When I'd written the first ad for them and showed it to Norman, he took one look at it and said, 'Jack, if you want to handle our advertising you better learn to fly." So I went out and took lessons. I learned to fly in a Mooney and have hardly ever flown any other kind of an airplane.

I remember one time, I had just gotten my license and didn't know much, when Hal Rachel was going to fly from Midland back to Kerrville. He asked me to come along. He was the darndest pilot. I had heard stories, but I'd never flown with him.

"We got out to the airport around 6 in the morning. It was still dark out and colder than you-know-what and there was driving snow. We took off and he kept saying to me, 'Keep looking at the wing-tip lights for ice.' There was no ice and after a while we broke out on top. Meanwhile, the sun starts to come up and all that weather is below us. First time I'd been VFR on top.

"He was the kind of pilot who'd just fly in a certain direction for as long as he figured it should take him to get to where he wanted to go and then he'd start looking. Well, when he figured that we must be somewhere around Kerrville, he said, 'I'm going down and you look out of the window and tell me when you see the ground. I'm going as low as 2400 feet. If you haven't seen the ground, we'll go back up.'

"So we let down into that stuff and the minute we got into it, bang, we got all iced up. The windshield was solid. Of course, I could still look out of the side, so we went down and down and down to 2400 feet and there was no land in sight. Back up we go, break out and the sun hits us and the ice melts off and he says, 'Let's try it again.'

"We do this three times. By this time, I'm getting pretty nervous, so I sort of mention to Hal that maybe we should go to San Antonio. So far Hal hadn't talked to anybody on the radio. But we'd listened to the weather. And the front, which was the reason for all that stuff, was by now halfway between San Antonio and Kerrville. So he flies right up to the edge of the front and drops down real low and turns back toward Kerrville along the road. He had me read the highway signs. We were that low. I remember seeing one billboard that said 11 miles to Kerrville. But just about that time, the clouds and the road came together, so back up through this crud we go.

"By now we'd been fooling around so long that the front had gotten to San Antonio. But he decided we better go to San Antonio anyway. So, still without talking to anybody, we fly to San Antonio and as we pass over the omni he says, 'Swell, there's a hole,' and we start corkscrewing down through this hole which wasn't much bigger than the airplane. 'Ah, there's the airport right there!' We were right on top of what he thought was San Antonio International. He called International Tower and was cleared to land. Right then I look and all of a sudden I see this name on top of one of the hangars: RANDOLPH. We had come down right on top of Randolph Air Force Base. So he, calm as anything, called the tower again and got vectors to International.

"Norm was another one of those casual pilots. If you asked him what would be the best route to, say, Miami, he'd say, 'Take up a heading of 90 degrees and when you see the ocean, turn right.' But he could handle an airplane like nobody else. In the beginning, when they were still working on the certification of the M-20, he'd fly around the country in a Mite. That was funny sight. He's a big man and he'd stand next to the airplane with his left hand inside the cockpit on the throttle, and with his right he'd swing the prop.

Cheaper Than Driving

"Until they brought out the M-20,Piper—with the Comanche, and Beech, with the Bonanza—had the high-performance, single-engine market sewn up. Cessna didn't have a rectractable single then and the thrust of our advertising was, 'Why be a wheel dragger. Pull the wheels up and go faster on less gas.' We made tests and we actually got 18 miles to a gallon, proving that you could fly cheaper than drive.

"By 1964, Mooney had captured a significant percentage of the market, largely at Piper's expense and that's when they started flying high. They started on that pressurized Mustang. The compnay, at the time, was making money and the theory was that they could build that thing for a million dollars out of cash flow.

"Piper, around that same time, conducted a survey and even had Ed Swearingen actually build a pressurized single-engine airplane for them. And they flew it, too, but decided that they were not going to go ahead with it because there wasn't enough of a market for it. But, in spite of that, Hal Rachel and Ralph Harmon decided to build.

Personally, I think that there very definitely would have been a market for it if they could build it cheap enough to sell for $29,000. I think the primary difference was that Piper was convinced that it couldn't be built for that kind of money, that it would cost more like $60,000, and at that price there was no market for it at that time. Ironically, as it turned out, $60,000 was what it was costing Mooney to build that airplane which they had promised to deliver at a profit for $29,000.

"Eventually their plans and goals just exceeded their financial capability and, in looking back, the down fall of Mooney started with the Mustang. That's where I fault Hal. That would have been the time to go public, but he thought that they'd make more money by not doing that. And then, when the money troubles started, it was too late. Just the same, I believe that in everything that Hal did he was sincerely trying to save that company. I think that he and Norman made a lot of personal sacrifices, and if it had all worked out, they'd both have ended up heroes."

Appendix A

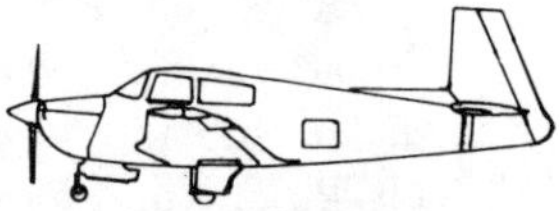

A Chronology Of The Mooney Aircraft Corporation

1946. Mooney Aircraft Inc. was formed by Al Mooney in partnership with Charles G. Yankey. It was located in Wichita, Kansas.

1952. The company moved to Kerrville, Texas, where it built, flight tested and certificated the Mark 20 (M-20).

1953. Charles G. Yankey died, leaving his 50 percent of stock in the company to Al Mooney. Hal Rachel, Norman Hoffman and Ed Hunnicut bought the company, receiving 100 percent of the stock.

1955. Al Mooney left the company and joined Lockheed.

1960. Ralph Harmon joined Mooney as chief design engineer.

1963. Negotiations with Mitsubishi Heavy Industries of Japan were started. This eventually led to an agreement to have Mooney assemble and market the MU-2 corporate turboprop.

1969. In order to facilitate the sale and refinancing of the company, Hal Rachel, Norm Hoffman and Ed Hunnicut resigned. Ralph Harmon became president. The company was then bought by American Electronics Laboratories of Pennsylvania. Being either unwilling or unable to shoulder the financial burden of revitalizing the company, AEL decided to opt for bankruptcy.

1970. Butler International Inc. of New Jersey acquired what was lef of the company. It also acquired the Aerostar and moved that program to Kerrville. No actual Aerostar twins were ever manufactured under the Butler banner, but Mooney Rangers, Chaparrals and Executives were briefly marked as single-engine Aerostars.

1971. Butler closed down production in Kerrville.

1973. Republic Steel bought the dormant company, reorganized it as Mooney aircraft Corp., a wholly owned subsidiary, and started production again.

1974. The first Mooneys manufactured by the new company came on the market .

1977. The 201 was introduced .

1979. The Turbo 231 was introduced .

1984. The estimated date of the introduction of the pressurized M-X.

Appendix B

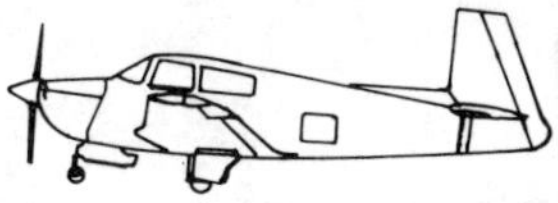

Illustrations and tables referenced in chapter 10, pages 92 - 98

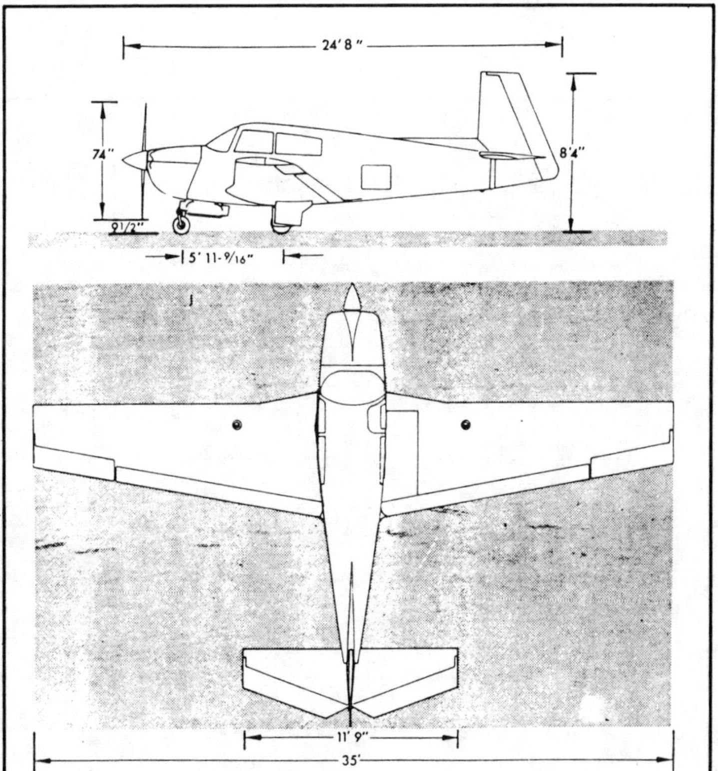

Fig. 10-1A. Two-views of the Mooney 201.

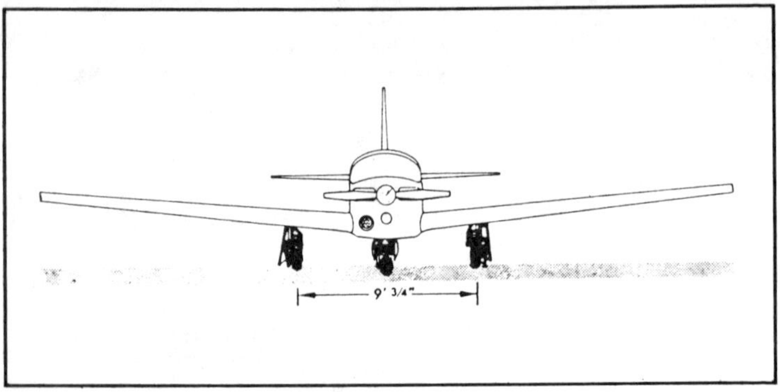

Fig. 10-1B. Third view of Mooney 201.

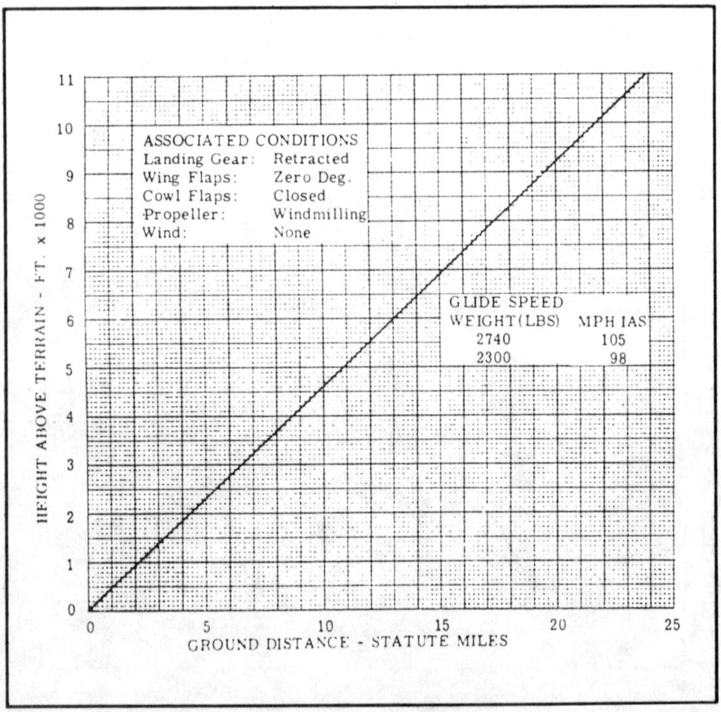

Fig. 10-2. Maximum glide distance model M20J.

$$TAS = CAS \times 1/\sqrt{\sigma}$$

EXAMPLE:

Given: 26°C Outside Air Temperature
3000 Ft. Pressure Altitude
162 Knot CAS

Find: 4950 Ft. Density Alt.
$1.077 = 1/\sqrt{\sigma}$
TAS = 162 x 1.077 = 174.5 Knots

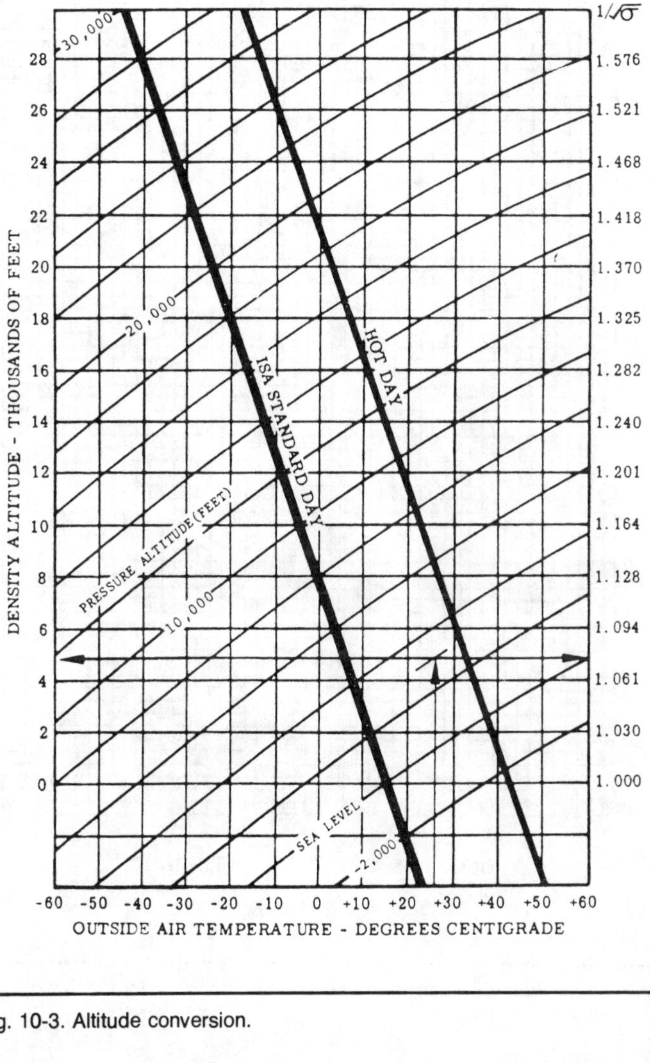

Fig. 10-3. Altitude conversion.

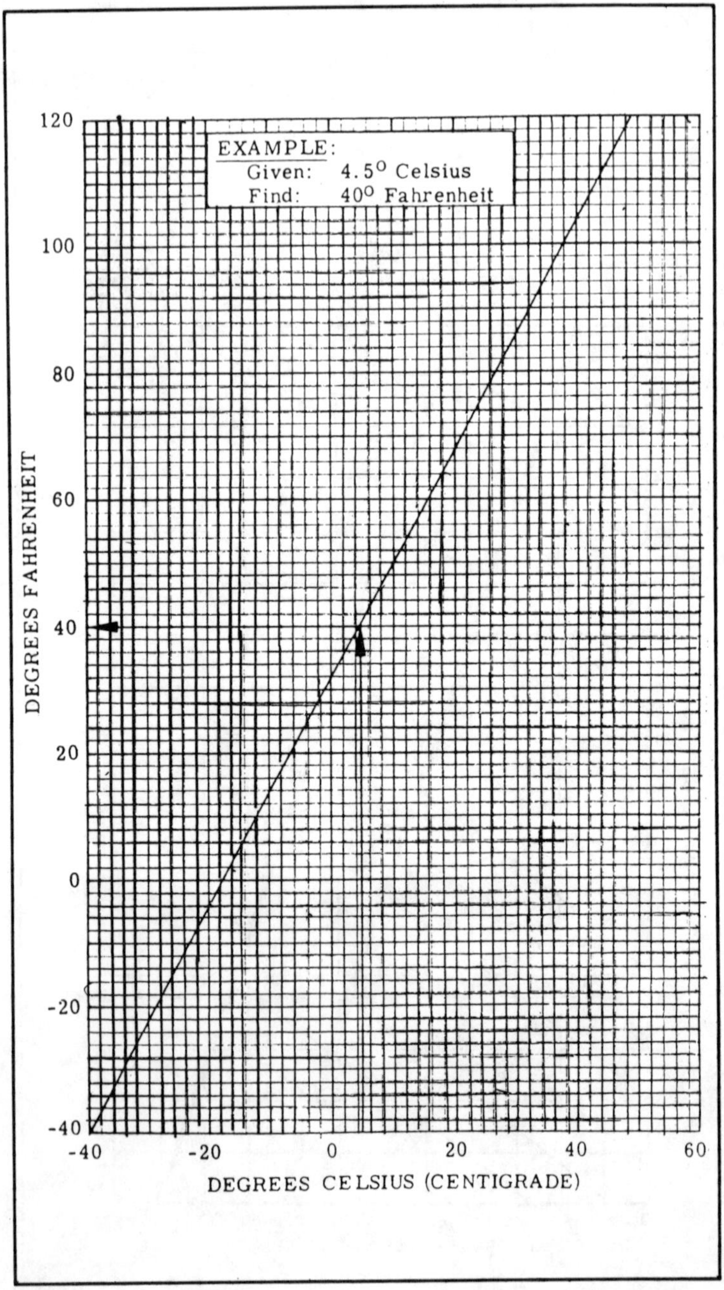

Fig. 10-4. The chart is an easy way to convert degrees centigrade to degrees Fahrenheit.

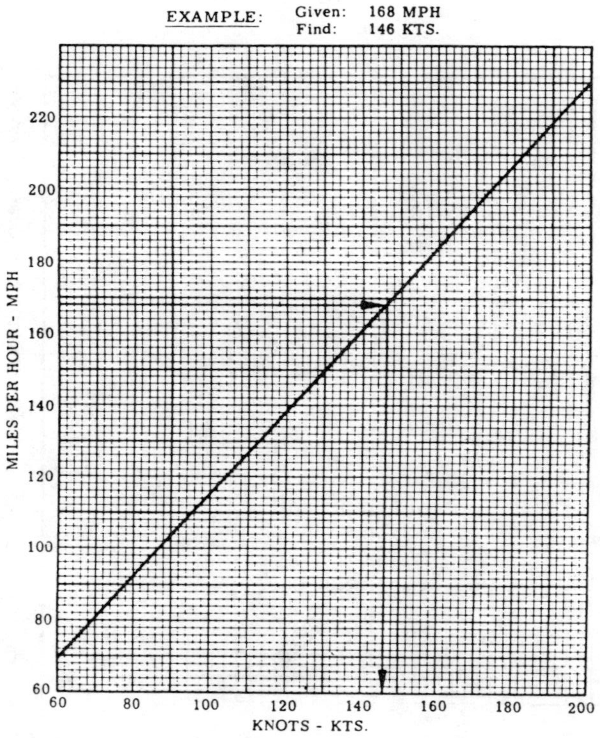

Fig. 10-5. The chart is used to convert mph to knots or vice versa. An alternate method is to multiply knots by1.15 or to divide mph by1.15.

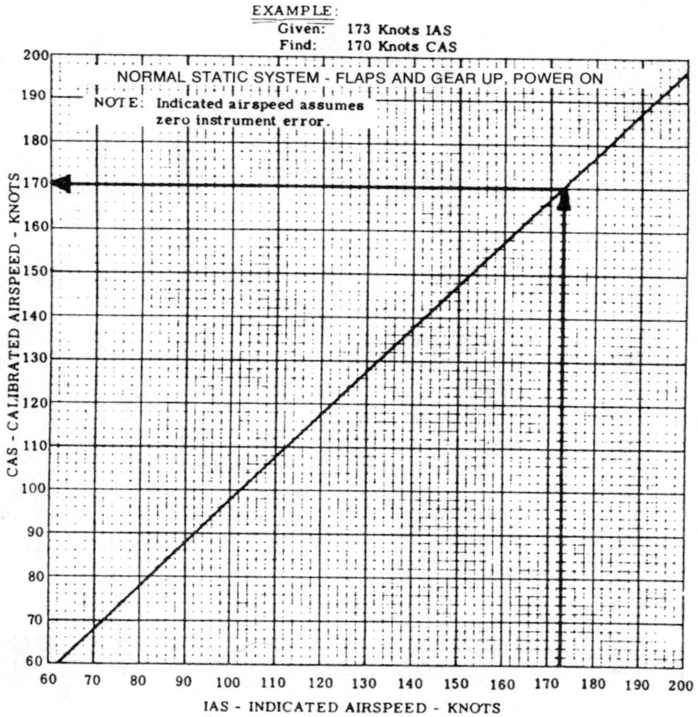

Fig. 10-6. This chart converts CAS to IAS or vice versa—with gear and flaps up and power on.

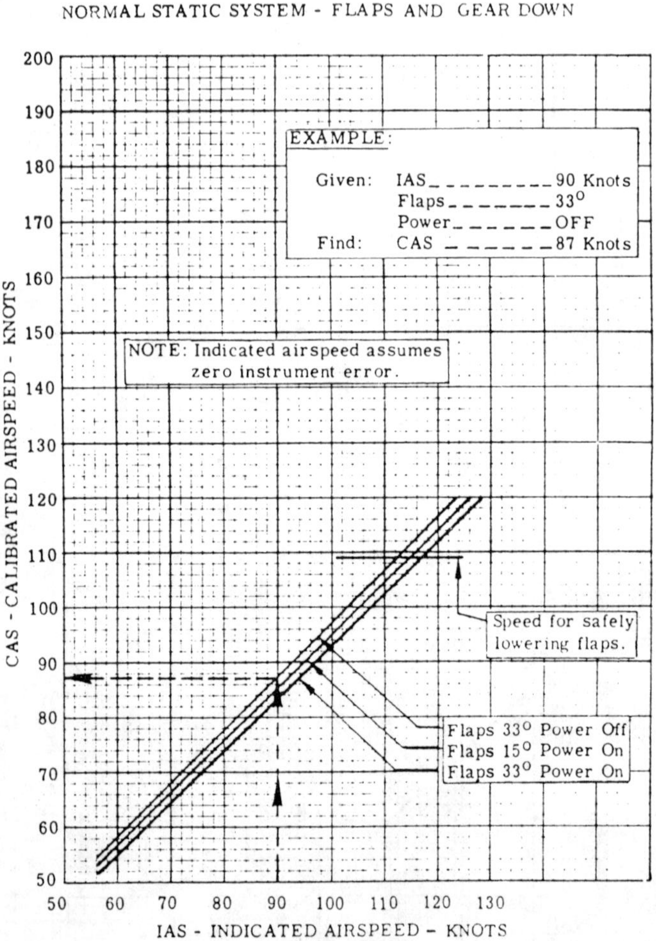

Fig. 10-7. Converting CAS to LAS while assuming that the gear is down and several different flap positions.

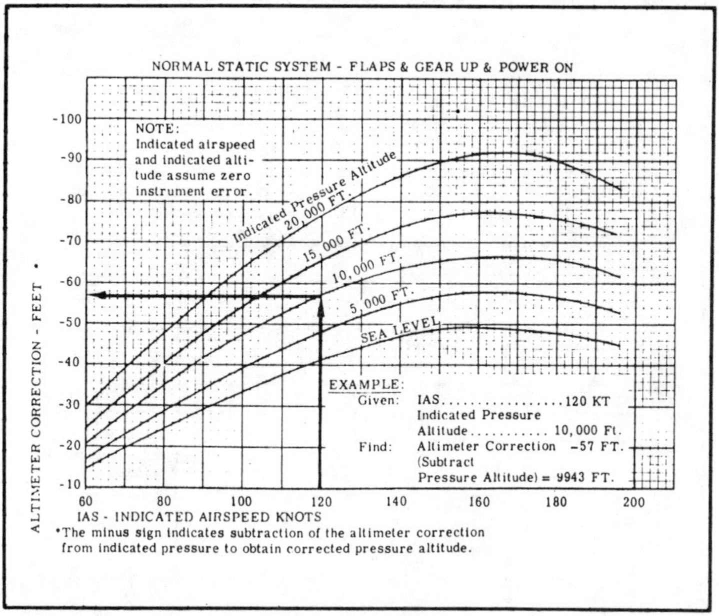

Fig. 10-8. Altimeter errors shown with gear and flaps up and power on.

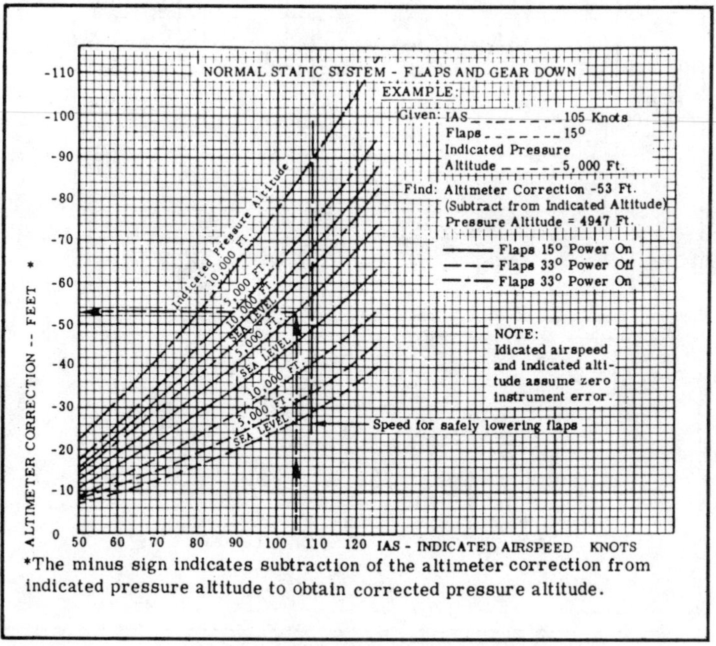

Fig. 10-9. The same with the gear down and different flap positions.

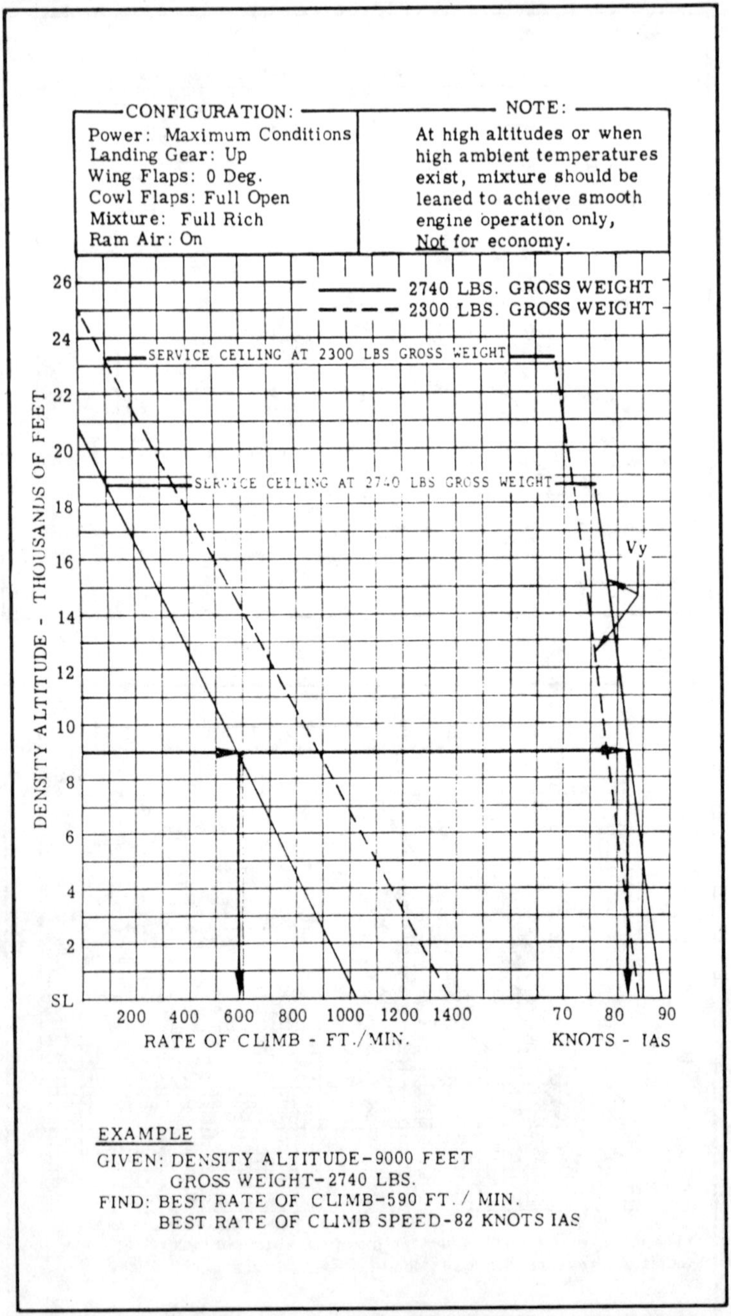

Fig. 10-10. The rate of climb under different density-altitude conditions.

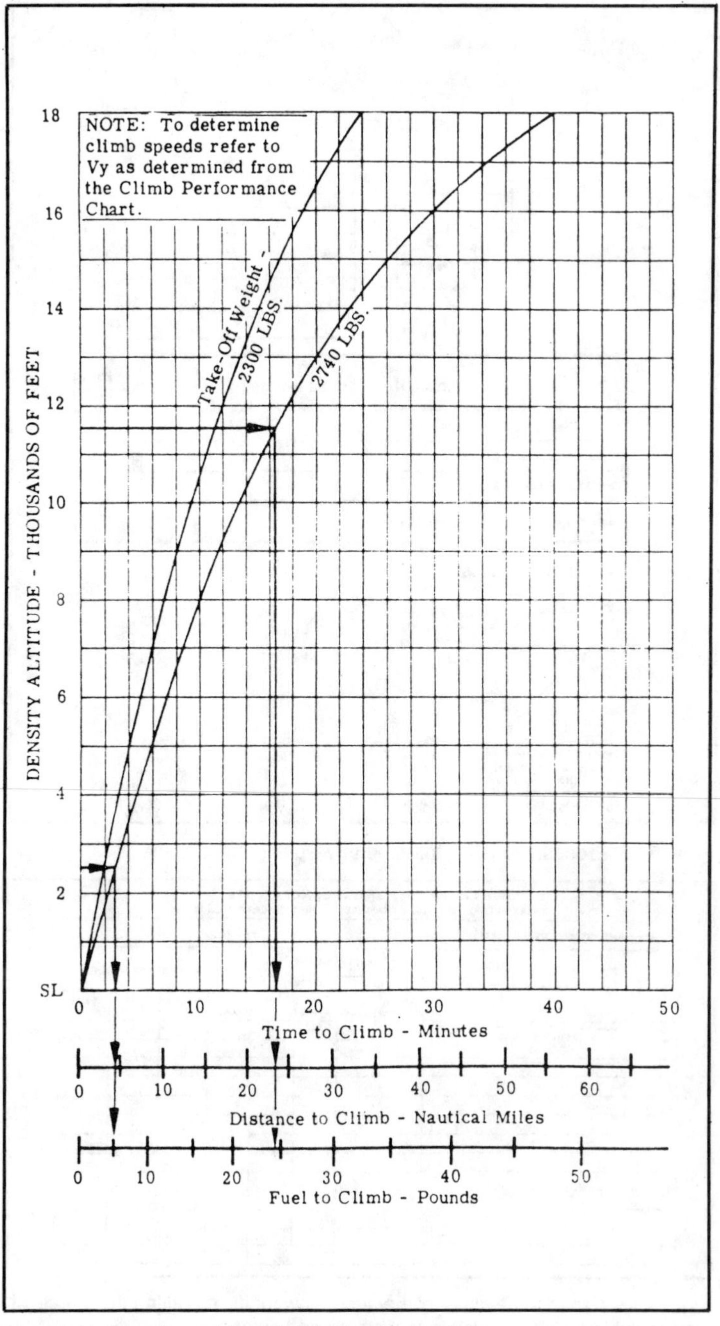

Fig. 10-11. The elapsed time, fuel burned and distance covered during climb.

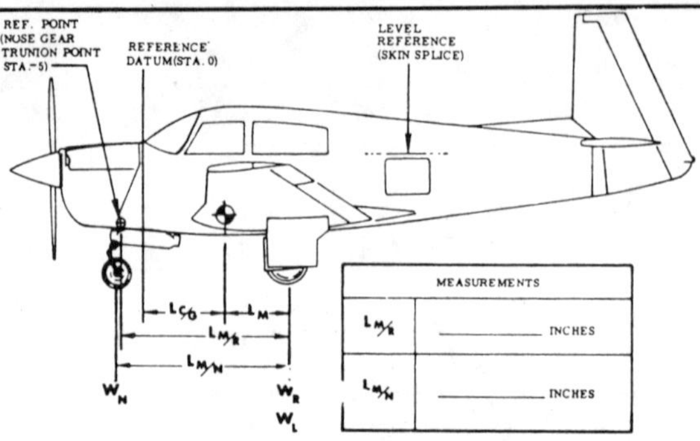

SCALE POSITION AND SYMBOL	SCALE READING	TARE	NET WEIGHT
Nose Wheel (W_N)			
Right Main Wheel (W_R)			
Left Main Wheel (W_L)			
Basic Empty Weight, as Weighed (W_T)			

a. CG Forward of Main Wheels:

______ LBS. Weight of Nose (W_N) X ______ IN. Distance Between Main and Nose Wheel Axle Centers ($L_{M/N}$) ÷ ______ LBS. Total Weight of Aircraft (W_T) = ______ IN. CG Forward of Main Wheels (L_M)

b. CG Aft of Datum (Station 0):

______ IN. Distance from Center Nose Gear Trunion to Center of Main Wheel Axles (Horizontal) ($L_{M/R}$) ______ IN. Distance from Nose Gear Trunion to Datum Constant − ______ IN. Result of Computation Above (L_M) = ______ IN. CG (FUS. STA.) Distance Aft of Datum. (Empty Weight CG) (L_{CG})

Fig. 10-12. Procedures to be followed in weighing the airplane.

OWNERS WEIGHT AND BALANCE RECORD
(ENTER BELOW ALL WEIGHT CHANGE DATA FROM AIRCRAFT LOG BOOK)

AIRPLANE MODEL - M20J	SERIAL NUMBER ______	FAA REGISTRATION NO. ______

DATE	DESCRIPTION OF MODIFICATION	WEIGHT CHANGE				RUNNING EMPTY WEIGHT			
		ADDED (+)		REMOVED (-)					
		Wt. (Pounds)	Arm (Inches)	Wt. (Pounds)	Arm (Inches)	Wt. (Pounds)	Moment /1000	Arm (Inches)	Useful Load
	BASIC EMPTY WEIGHT AS DELIVERED (W_T) (INCLUDES FULL OIL - 8 QTS)	—	—	—	—				

Fig. 10-13. A form to be used by the aircraft owner in computing the effects of weight changes with reference to weight and balance.

PROBLEM FORM

FAA REGISTRATION NO. ______________ M20J SERIAL NO. __________

Step	ITEM	Sample Problem Pilot & Two Pass.		Your Problem	
		Weight (LBS)	Moment (LB-INS. /1000)	Weight (LBS)	Moment (LB-INS. /1000)
1	**Aircraft Basic Empty Weight, W_T (From Page 6-5) Includes Full Oil — 8 QT. @1.875 LBS/QT (Sta - 11.5) (Sump assumed full for all flights)**	**1710.0**	**75.26**		
2	Pilot Seat (#1)*	170.0	6.0 (2nd Pos.)		
	Copilot Seat (#2)*	170.0	5.8 (Fwd. Pos.)		
3	Left Rear Seat (#3)	170.0	12.00		
	Right Rear Seat (#4)				
4	Fuel (Max. Usable 64 Gal. 384 LBS @ sta 48.43)	312.0	15.11		
5	Baggage (Max. 120 LBS @ Sta 95.5)	110.0	10.23		
	Hat Rack (Max. 10 LBS @ Sta 119.0)	3.0	.36		
6	Loaded Aircraft Weight	2645.0	X		X
	Total Moment/1000	X	124.76	X	
7	Refer to Page 6-8, Center-of-Gravity Moment Envelope, to determine whether your aircraft loading is acceptable.				

*Obtain the moment/1000 value for each seat position (FWD, MID, or AFT.) from loading computation graph below.

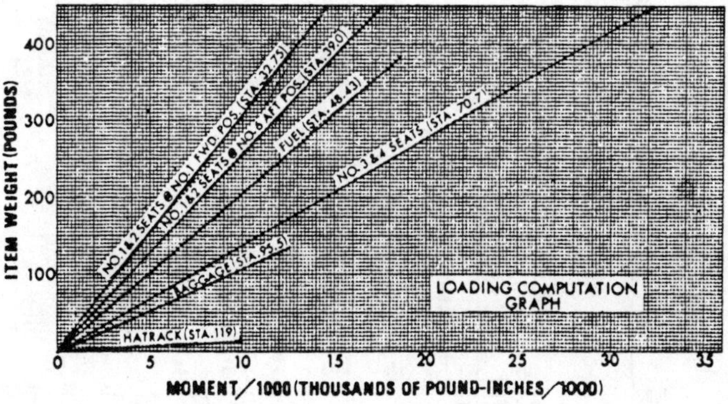

Fig. 10-14. Weight-and-balance problem form and loading computation graph.

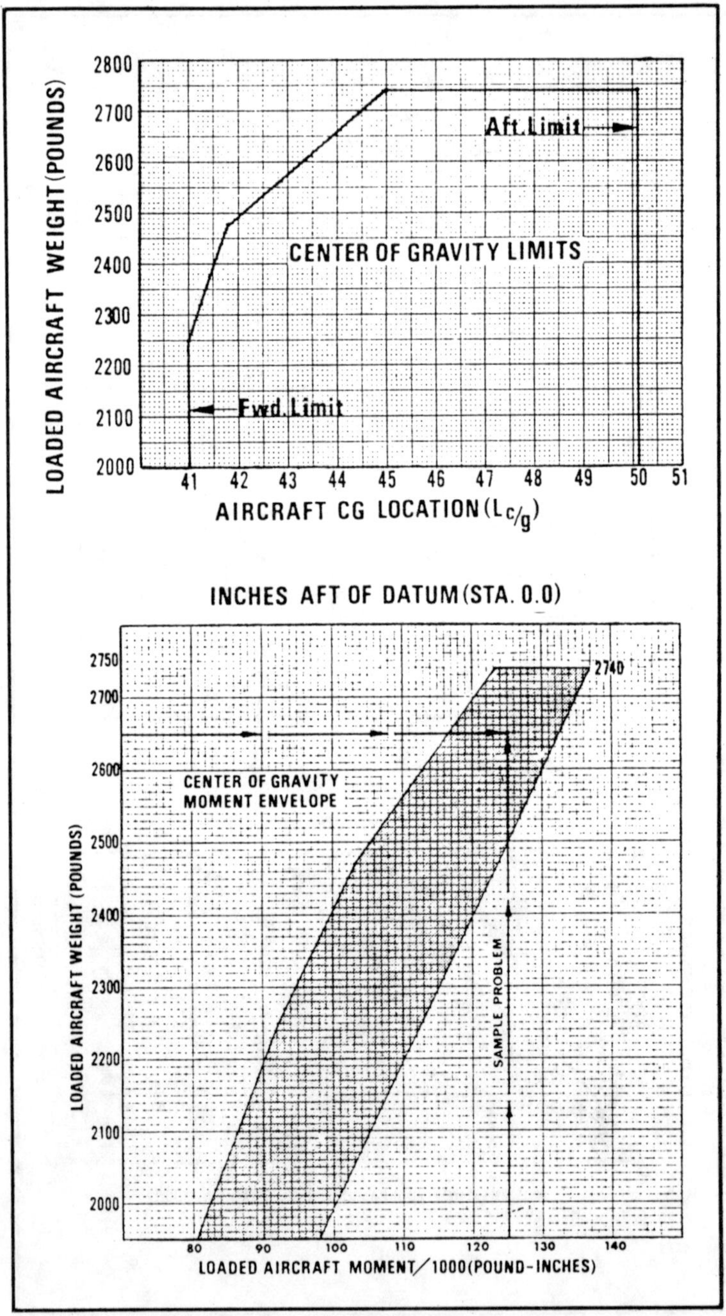

Fig. 10-15. Center-of-gravity moment envelope and cg limits.

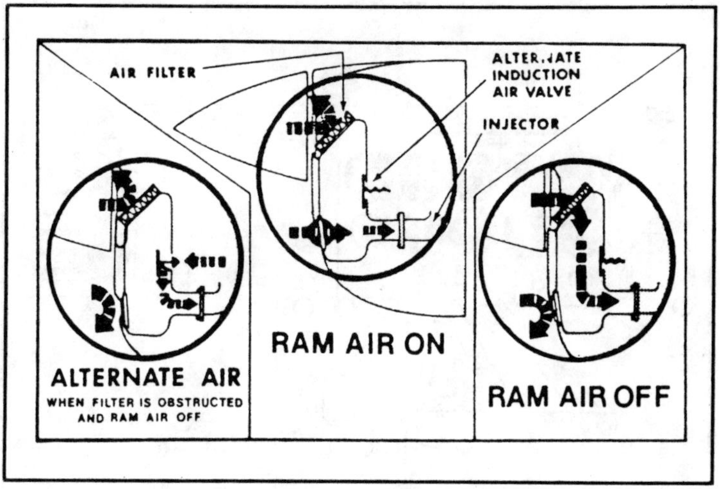

Fig. 10-16. The engine-air induction system.

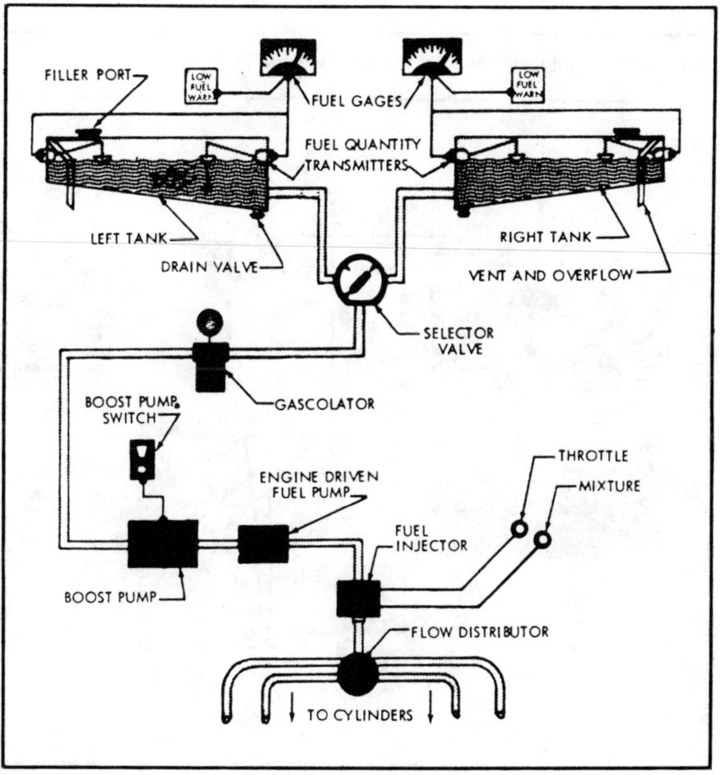

Fig. 10-17. The fuel system.

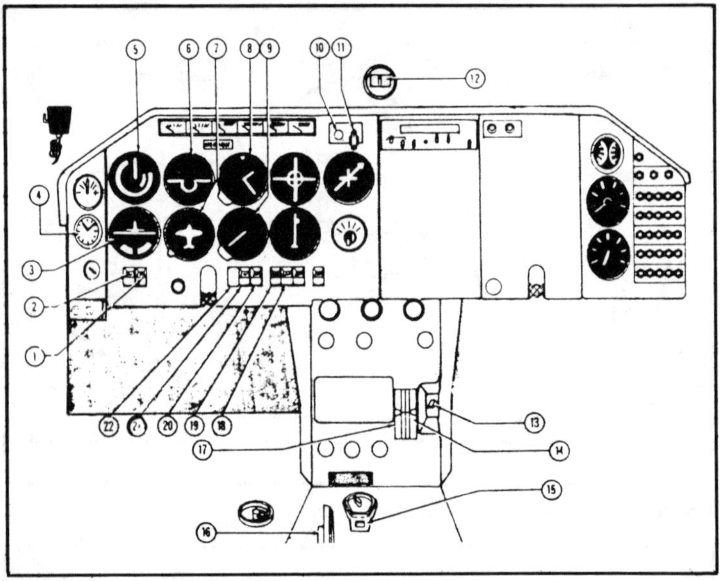

Fig. 10-18. Flight instruments and controls of the 201.

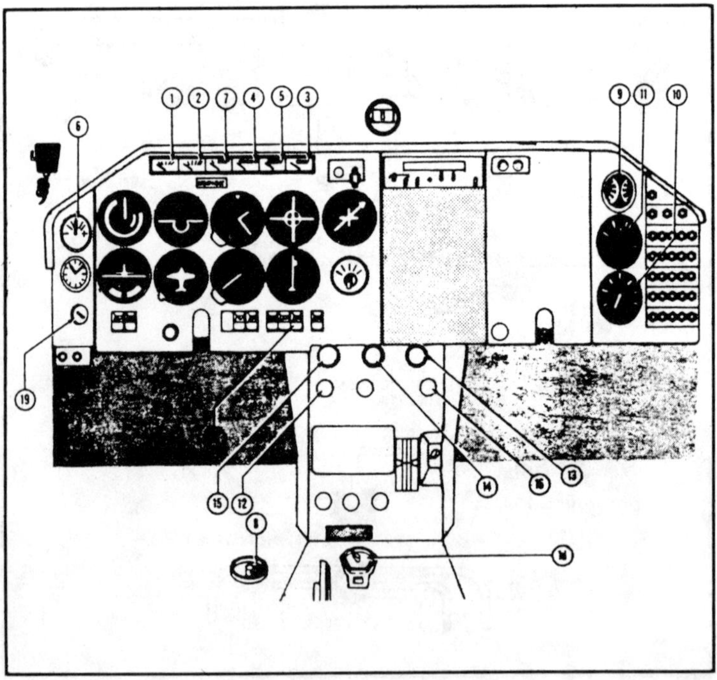

Fig. 10-19. Engine instruments and controls for the 201.

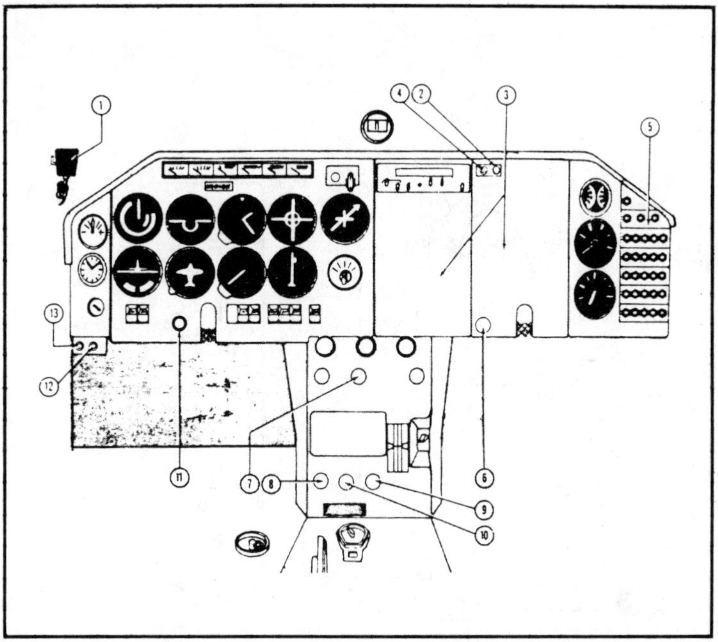

Fig. 10-20. Miscellaneous instruments, controls and indicators on the 201.

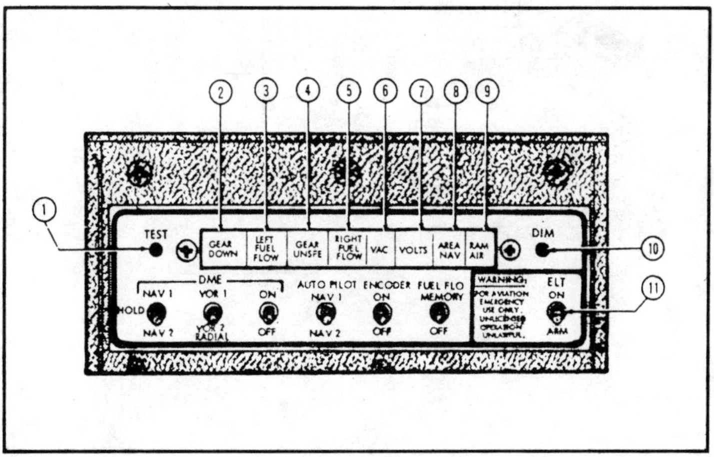

Fig. 10-21. Annunicator and switch panel of the 201.

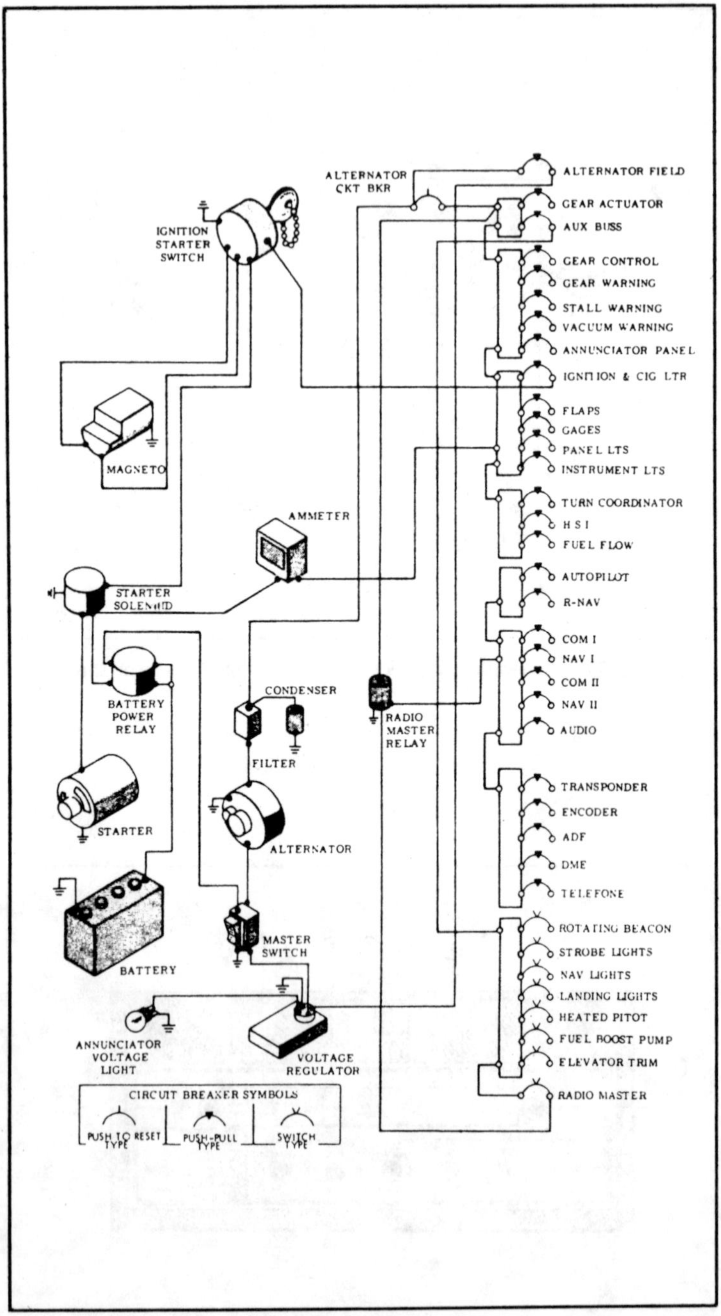

Fig. 10-22. Diagram of the electrical system of the 201.

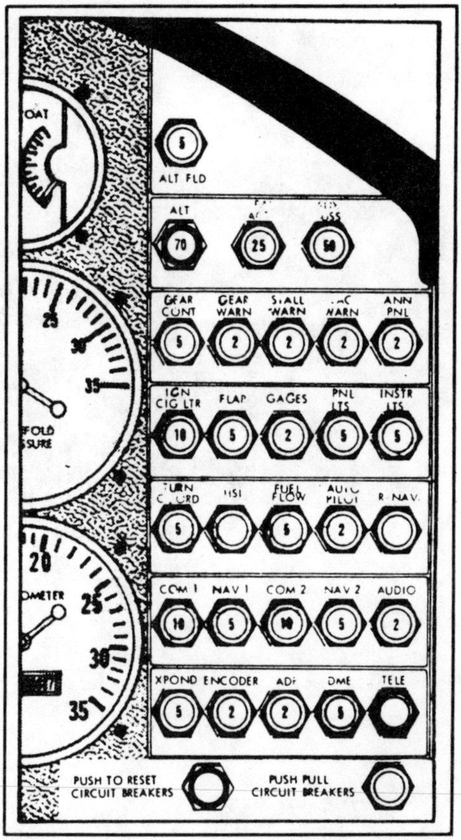

Fig. 10-23. Circuit-breaker panel in the 201.

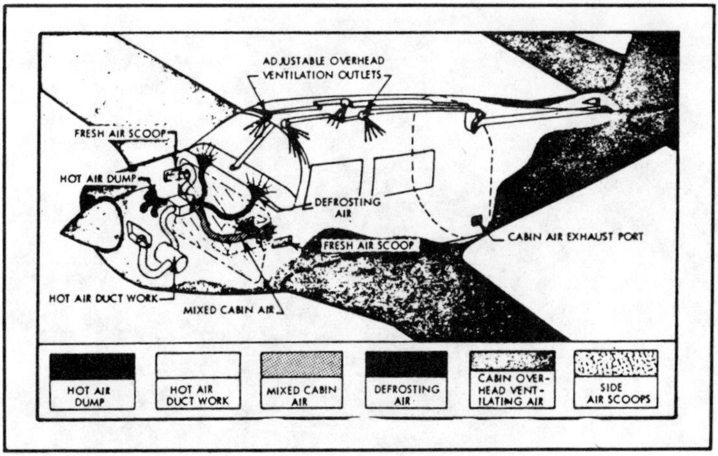

Fig. 10-24. The heating and ventilation system of the 201.

Fig. 10-25. The instrument panel of the 201.

Fig. 10-26. Part of the radio stack topped by the annunciator panel in the 201.

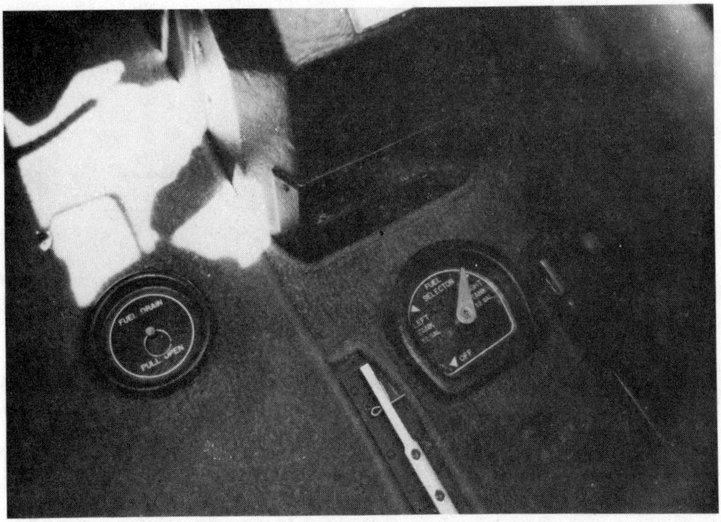

Fig. 10-27. The fuel strain and switching units in the floor in front of the left seat of the 201.

Fig. 10-28. Using conventional throttle, mixture and rpm controls rather than the now-popular pedestal type adds room in the cokpit and permits better control.

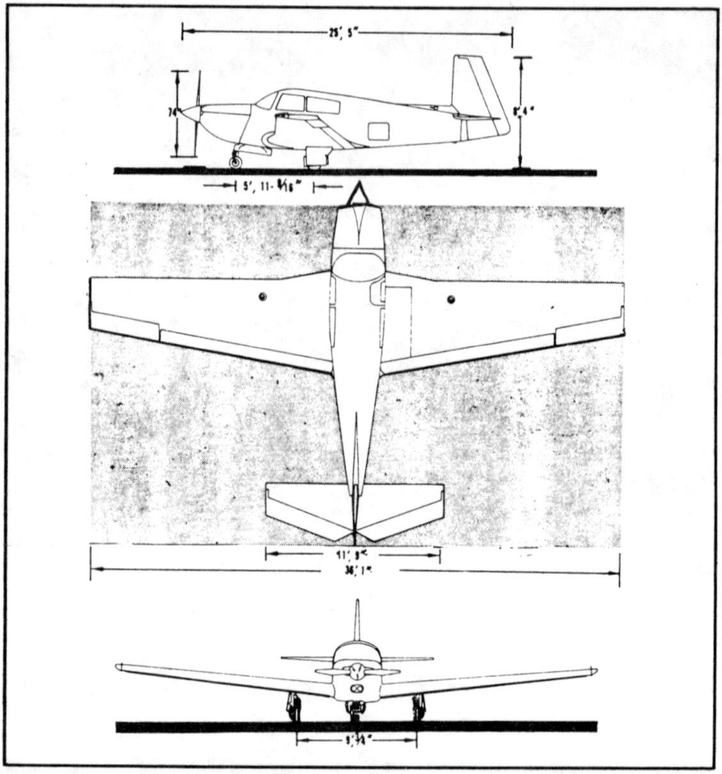

Fig. 10-29. Three-views of the Mooney Turbo 231.

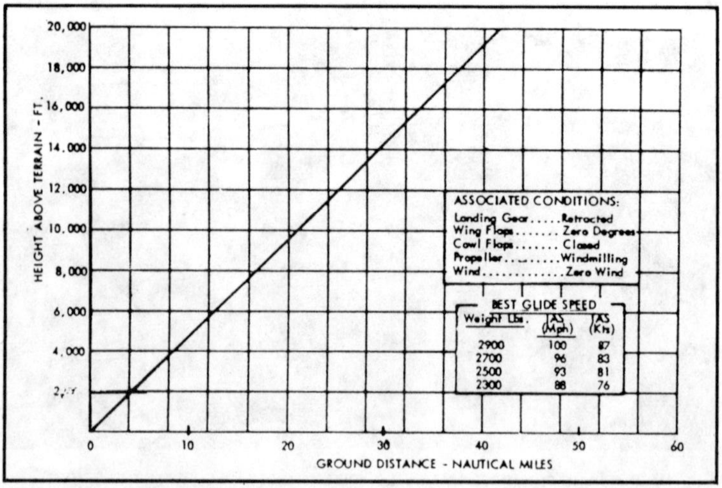

Fig. 10-30. Glide distances for the 231.

TAS = CAS x $1/\sqrt{\sigma}$

EXAMPLE:

Given: 26°C Outside Air Temperature
3000 Ft. Pressure Altitude
162 Knot CAS

Find: 4950 Ft. Density Alt.
$1.077 = 1/\sqrt{\sigma}$
TAS = 162 x 1.077 = 174.5 Knots

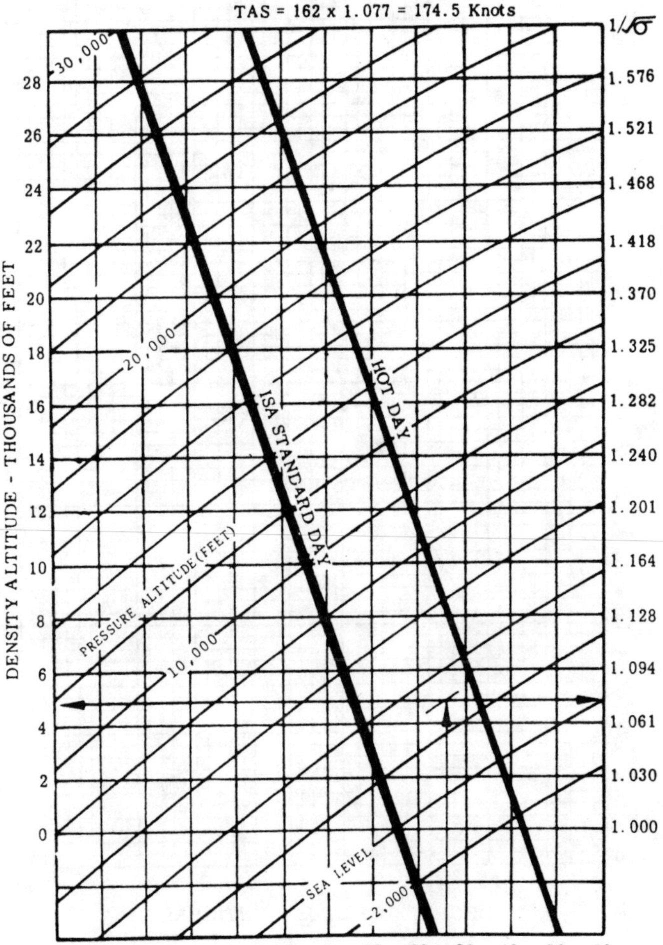

Fig. 10-31. Airspeed conversion chart TAS/CAS.

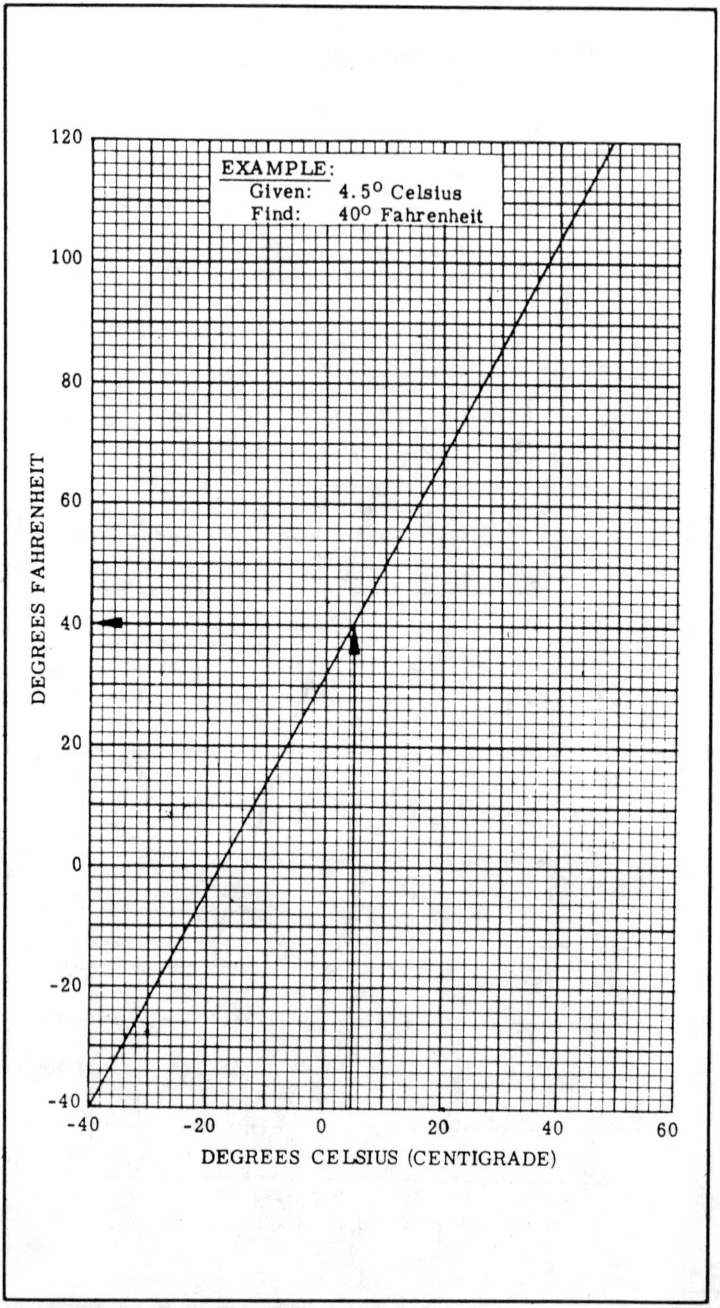

Fig. 10-32. Temperature conversion chart F./C.

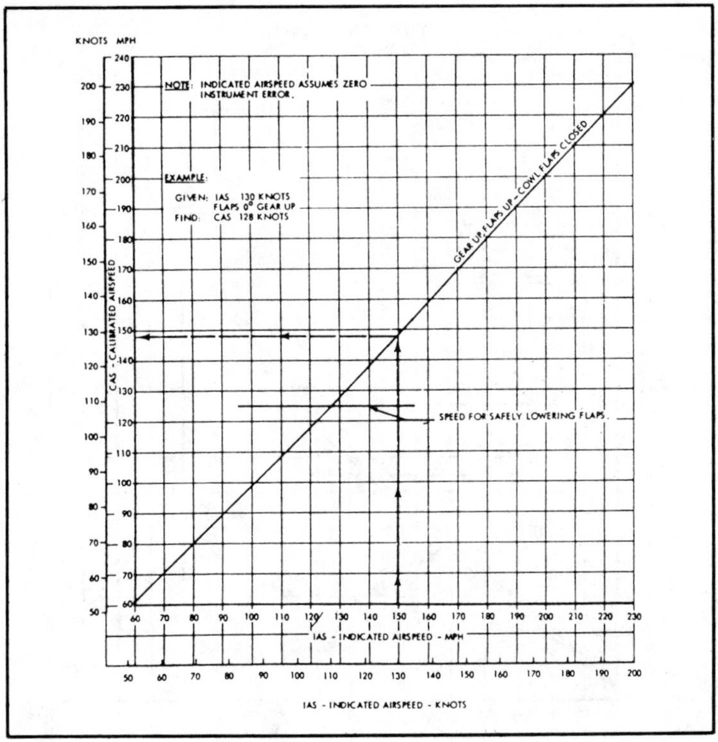

Fig. 10-33. Airspeed calibration with gear and flaps up.

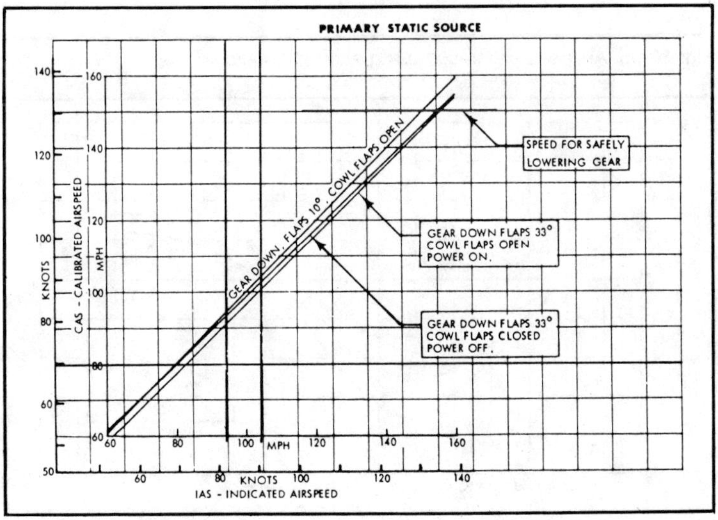

Fig. 10-34. Airspeed calibration with gear down and different flap settings.

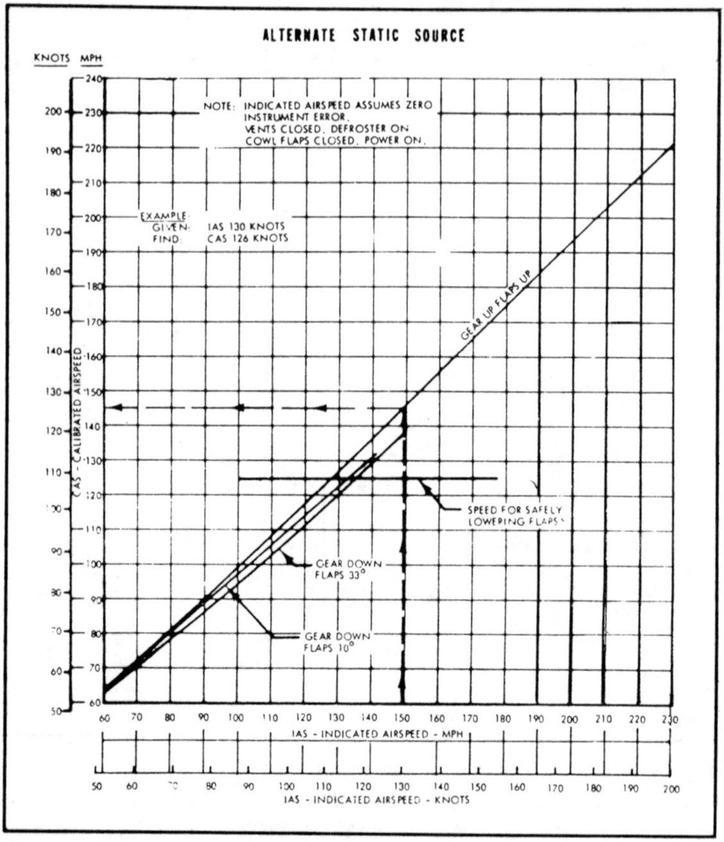

Fig. 10-35. Airspeed calibration using alternate static source.

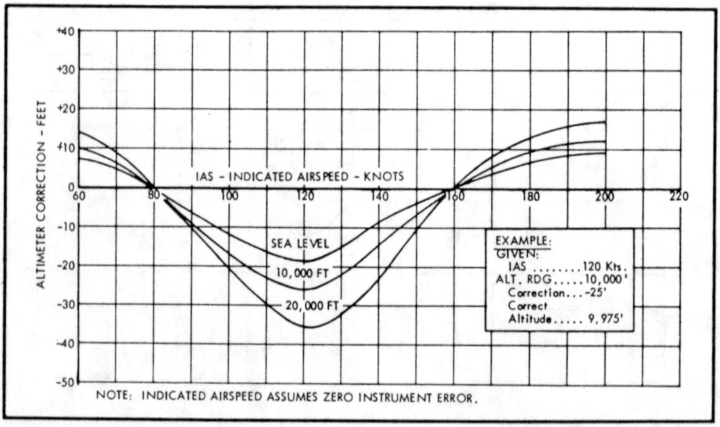

Fig. 10-36. Altimeter errors, gear and flaps up, using primary static source.

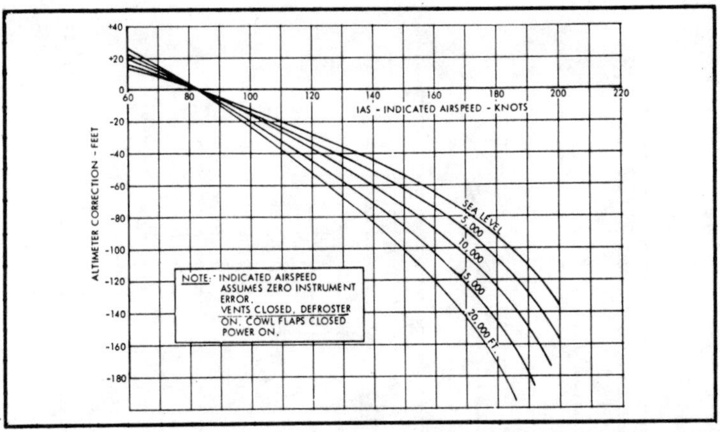

Fig. 10-37. Altimeter errors, gear and flaps up, using alternate static source.

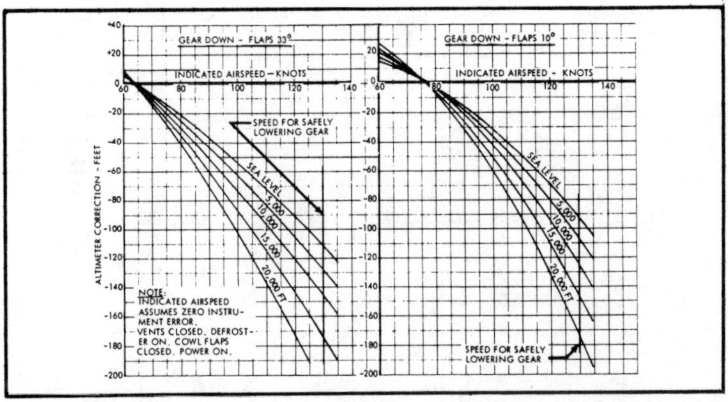

Fig. 10-38. Altimeter errors, gear down and two different flap positions, using alternate static source.

AFT C.G.

Gross Weight	Flap Deflection	Angle of Bank 0° CAS MPH (KTS)	0° IAS MPH (KTS)	30° CAS MPH (KTS)	30° IAS MPH (KTS)	45° CAS MPH (KTS)	45° IAS MPH (KTS)	60° CAS MPH (KTS)	60° IAS MPH (KTS)
2900 Lbs.	0°	71 (61.7)	68.5 (59.5)	76.3 (66.3)	75 (65.1)	84.4 (73.3)	84 (72.9)	100.4 (87.2)	101 (87.7)
	10°	70 (60.8)	68 (59.1)	75.2 (65.3)	74 (64.3)	83.2 (72.3)	82.5 (71.6)	99 (86)	99.5 (86.4)
	33°	64 (55.6)	63 (54.7)	67.7 (58.8)	66.7 (57.9)	76.1 (66.1)	76.1 (66.1)	90.5 (78.6)	97 (84.2)

FWD. C.G.

Gross Weight	Flap Deflection	Angle of Bank 0° CAS MPH (KTS)	0° IAS MPH (KTS)	30° CAS MPH (KTS)	30° IAS MPH (KTS)	45° CAS MPH (KTS)	45° IAS MPH (KTS)	60° CAS MPH (KTS)	60° IAS MPH (KTS)
2900 Lbs	0°	72 (62.5)	70 (60.8)	77.4 (67.2)	76 (66)	85.6 (74.3)	85 (73-8)	101.8 (88.4)	102.8 (89.3)
	10°	71 (61.7)	69.2 (60.1)	76.2 (66.2)	75 (65.1)	84.4 (73.3)	83.6 (72.6)	100.4 (87.2)	101 (87.7)
	33°	66 (57.3)	65 (56.4)	69.9 (60.7)	69.3 (60.2)	78.5 (68.2)	79 (68.6)	93.3 (81)	94.8 (82.3)

Fig. 10-39. Stall speeds at different angles of bank.

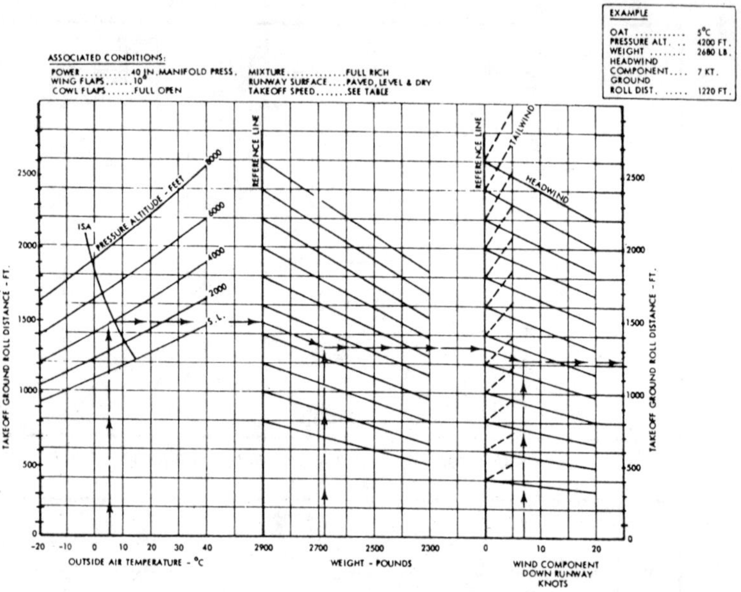

TAKEOFF WEIGHT LBS.	TAKEOFF SPEED MPH	(KTS) IAS
2900	77	67
2800	76	66
2700	74	65
2600	73	64
2500	72	63
2400	71	61
2300	70	60

Fig. 10-40. Takeoff ground roll under various conditions.

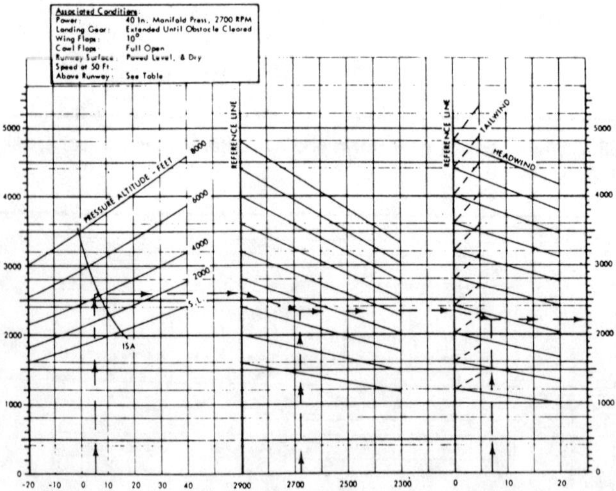

Takeoff Weight Lbs.	Speed At 50 Ft. MPH	(KTS) IAS
2900	85	74
2800	83	73
2700	82	71
2600	80	70
2500	79	69
2400	77	67
2300	75	65

EXAMPLE:

OAT: 5°C
Pressure Altitude: 4200 Ft.
Weight: 2680 Lb.
Headwind Component: 7 Kt.

Total Takeoff Distance Over A 50 FT. Obstacle: 2200 Ft.

Fig. 10-41. Takeoff distance over a 50-foot obstacle under various conditions.

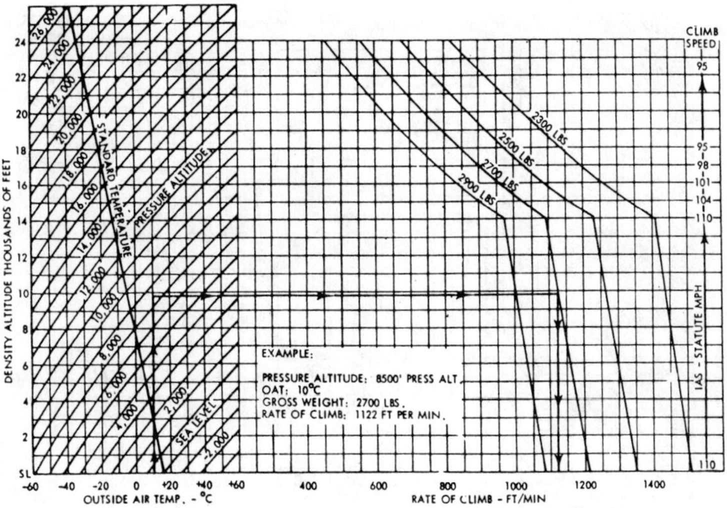

Fig. 10-42. Rate of climb under different load and temperature conditions.

Fig. 10-43. Elapsed time, fuel burned and distance covered during climb.

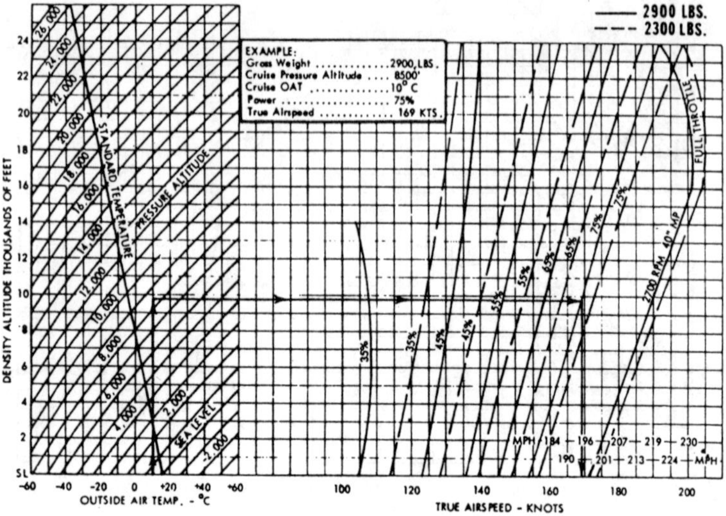

Fig. 10-44. Speed and power versus altitude at 2,900 and 2,300 pounds of aircraft weight.

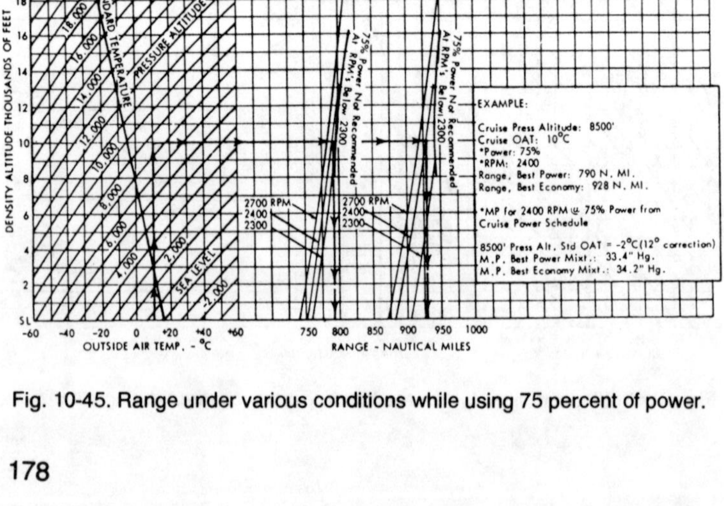

Fig. 10-45. Range under various conditions while using 75 percent of power.

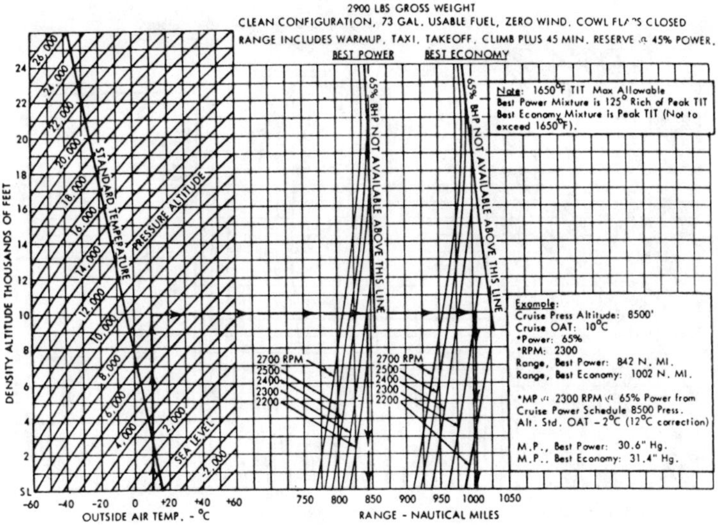

Fig. 10-46. Range under various conditions while using 65 percent of power.

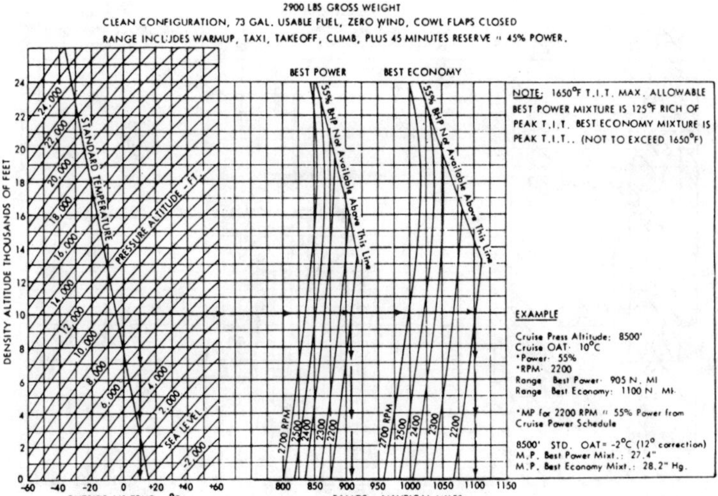

Fig. 10-47. Range under various conditions while using 55 percent of power.

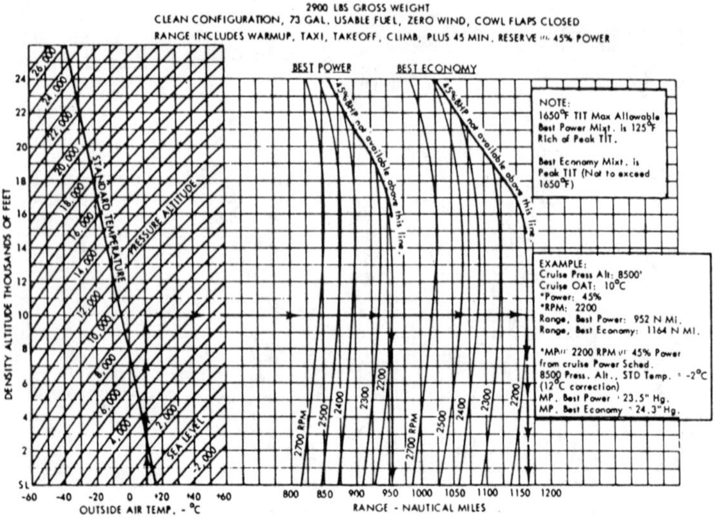

Fig 10-48. Range under various conditions while using 45 percent of power.

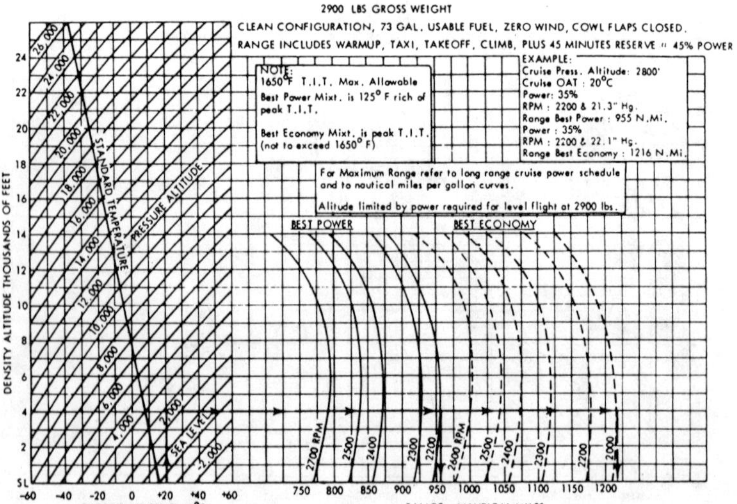

Fig. 10-49. Range under various conditions while using 35 percent of power.

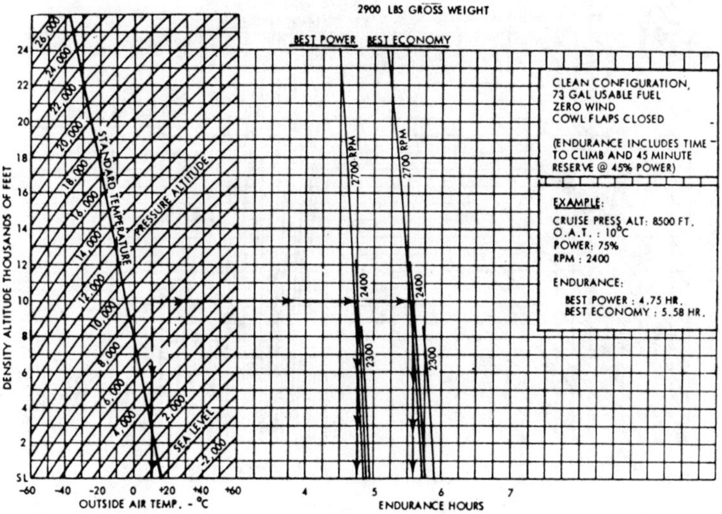

Fig. 10-50. Endurance under various conditions while using 75 percent of power.

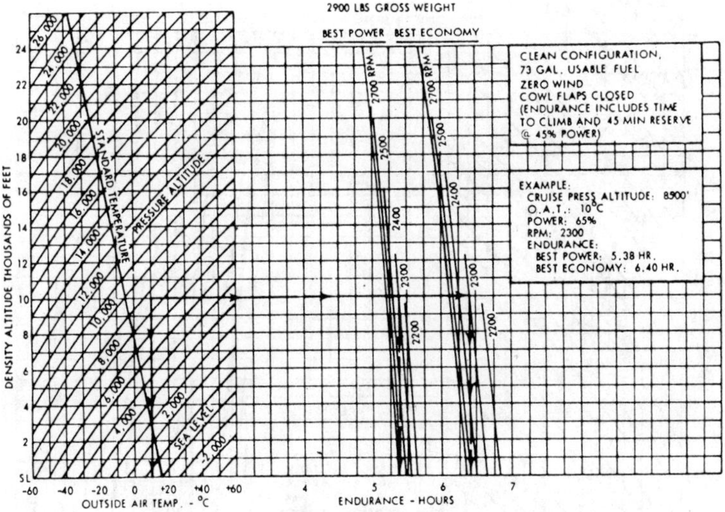

Fig. 10-51. Endurance under various conditions while using 65 percent of power.

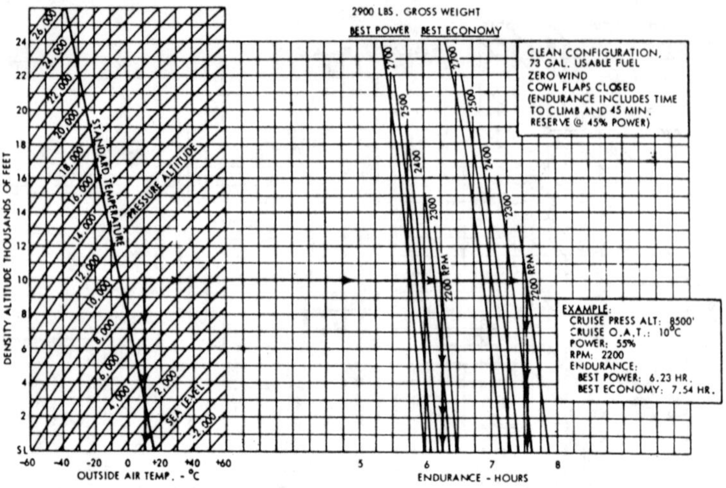

Fig. 10-52. Endurance under various conditions while using 55 percent of power.

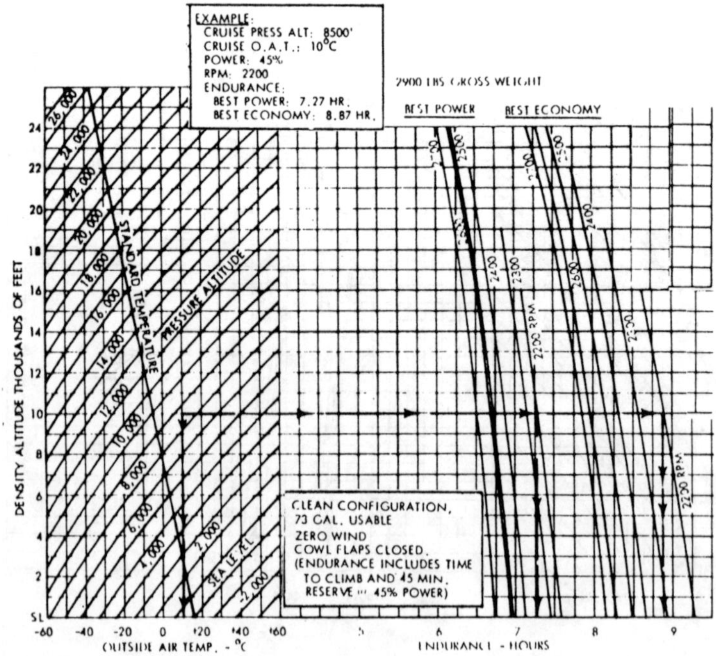

Fig. 10-53. Endurance under various conditions, using 45 percent of power.

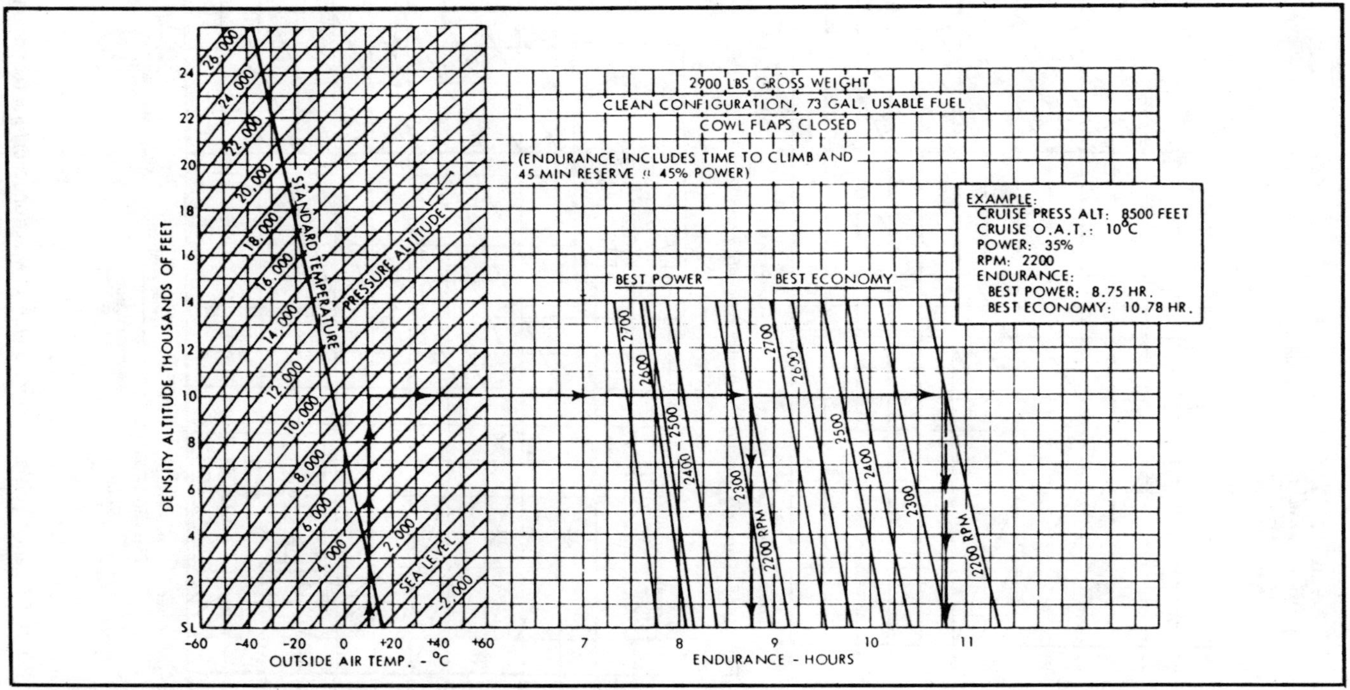

Fig. 10-54. Endurance under various conditions while using 35 percent of power.

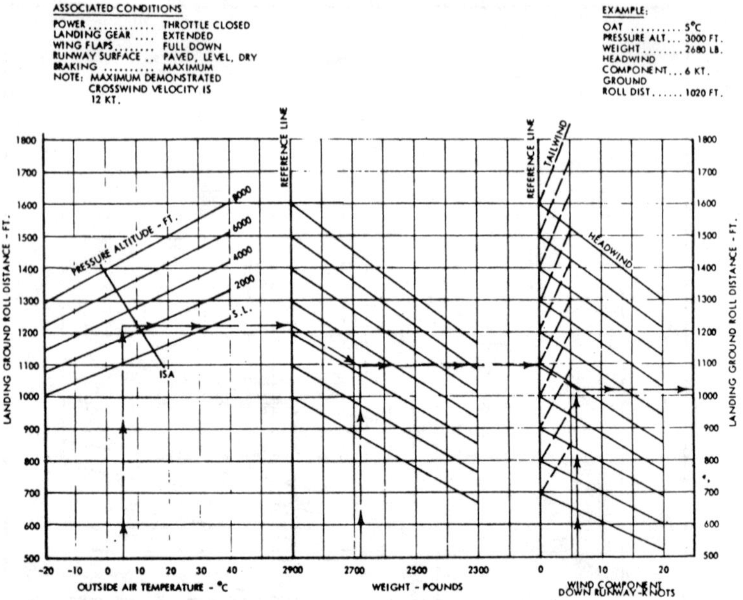

Fig. 10-55. Landing ground roll under different conditions.

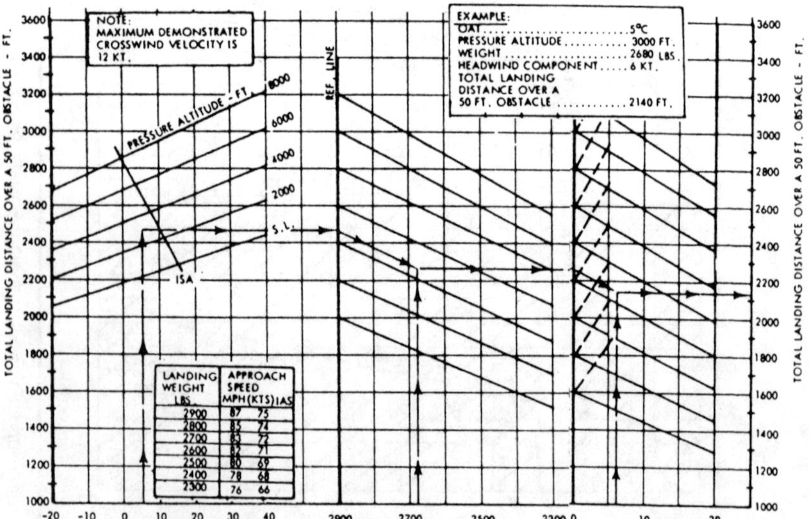

LANDING WEIGHT LBS.	APPROACH SPEED MPH	(KTS) IAS
2900	87	75
2800	85	74
2700	83	72
2600	82	71
2500	80	69
2400	78	68
2300	76	66

ASSOCIATED CONDITIONS:

POWER	:	THROTTLE CLOSED
LANDING GEAR	:	EXTENDED
WING FLAPS	:	FULL DOWN
APPROACH SPEED AT 50 FT.	:	SEE TABLE
RUNWAY SURFACE	:	PAVED, LEVEL, DRY
BRAKING	:	MAXIMUM

Fig. 10-56. Landing distance over a 50-foot obstacle under different conditions.

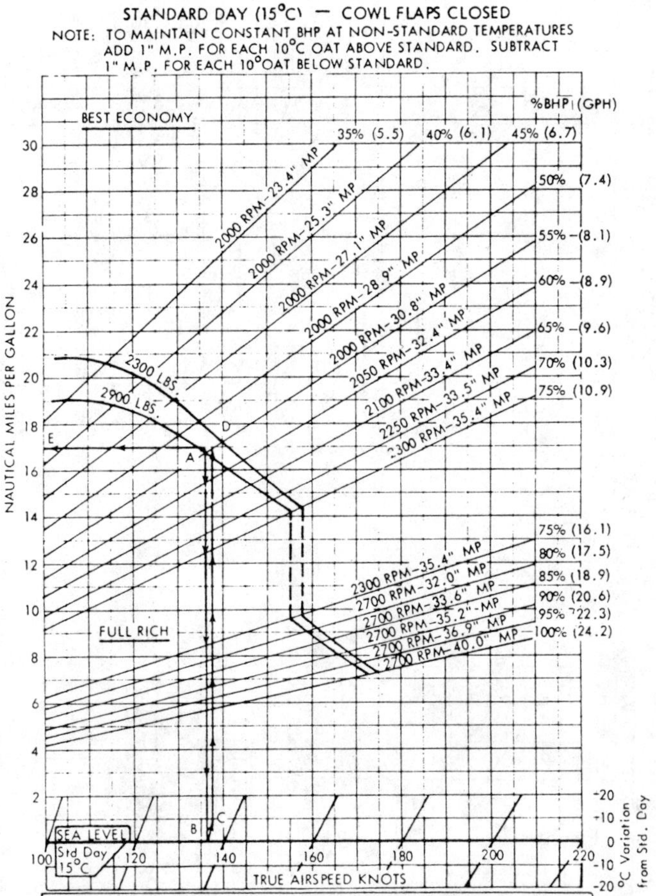

EXAMPLE: DO NOT EXCEED 1650° T.I.T.

GIVEN:	FIND:
Gross Weight....2900 Lbs.	Power............55%
Power..........55%	RPM............ 2000
RPM............2000	MP............ 31.8"
MP.............30.8"	GPH............ 8.1
OAT............25°C	N MI/GAL........ 17
GPH............8.1	True Airspeed...... 138 Kts.

Fig. 10-57. Nautical miles per gallon at long-range cruise power at sea level.

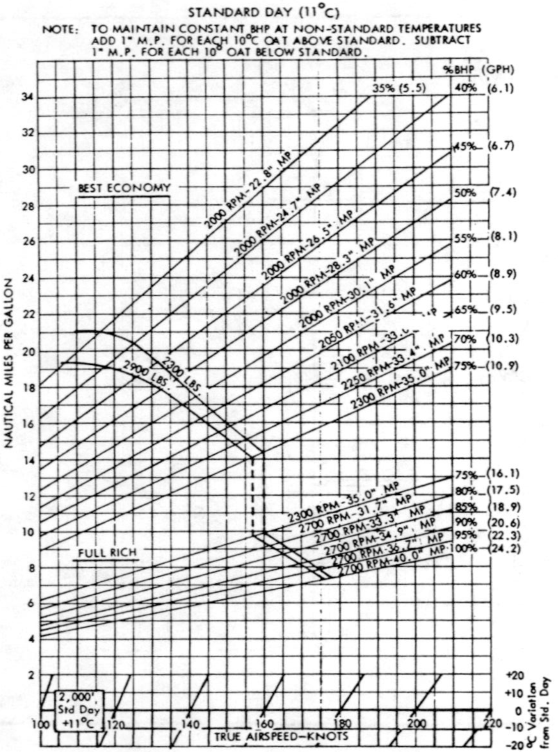

Fig. 10-58. Nautical miles per gallon at long-range cruise power at 2,000 feet.

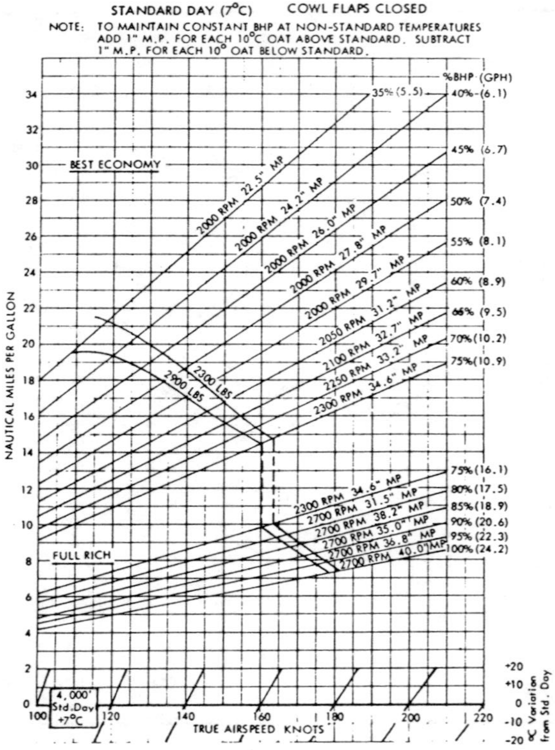

Fig. 10-59. Nautical miles per gallon at long-range cruise power at 4,000 feet.

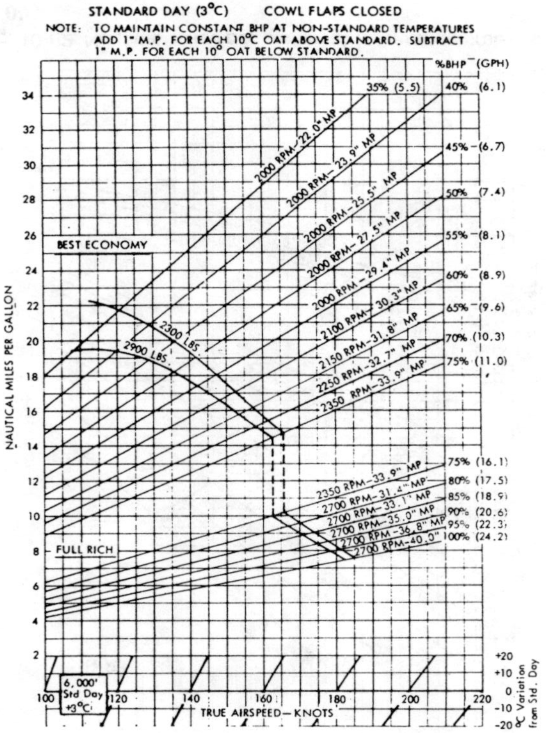

Fig. 10-60. Nautical miles per gallon at long-range cruise power at 6,000 feet.

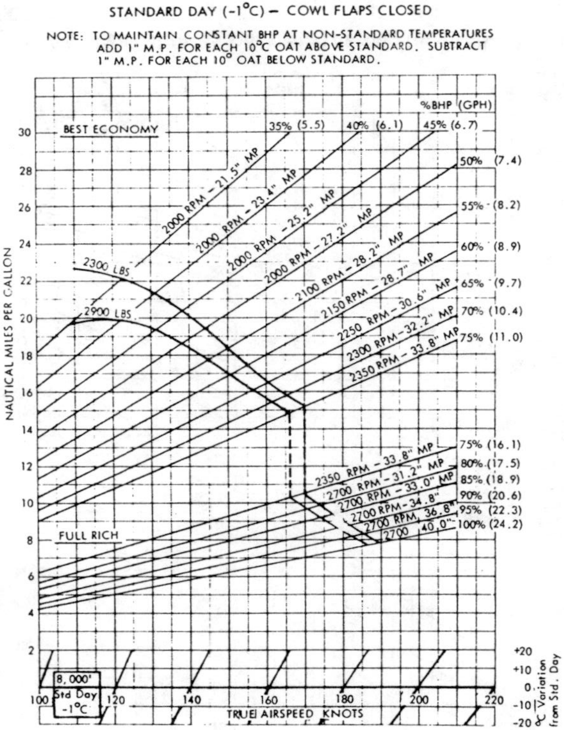

Fig. 10-61. Nautical miles per gallon at long-range cruise power at 8,000 feet.

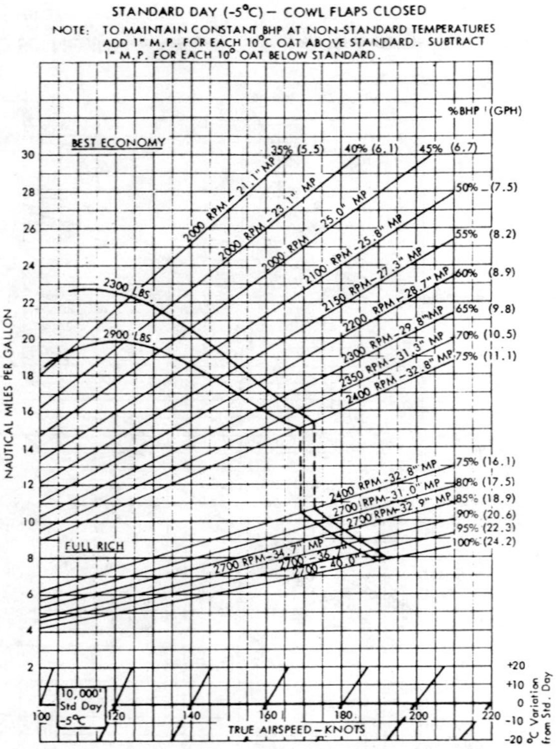

Fig. 10-62. Nautical miles per gallon at long-range cruise power at 10,000 feet.

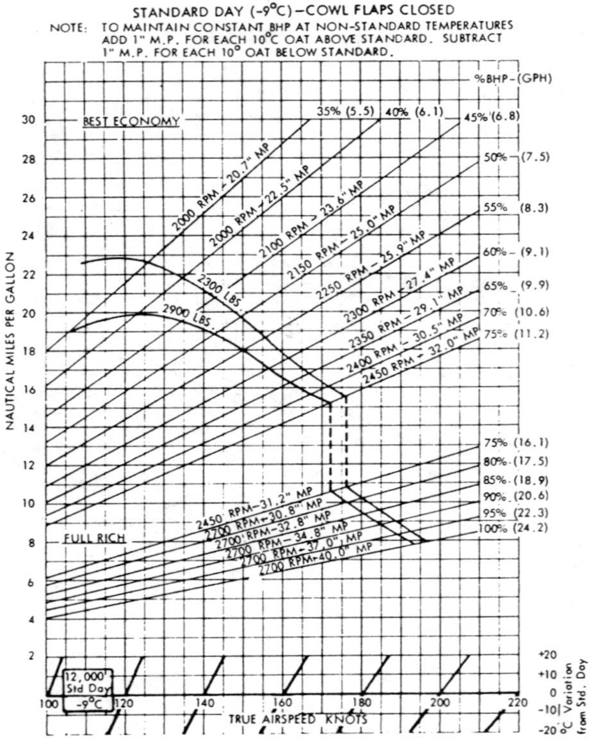

Fig. 10-63. Nautical miles per gallon at long-range cruise power at 12,000 feet.

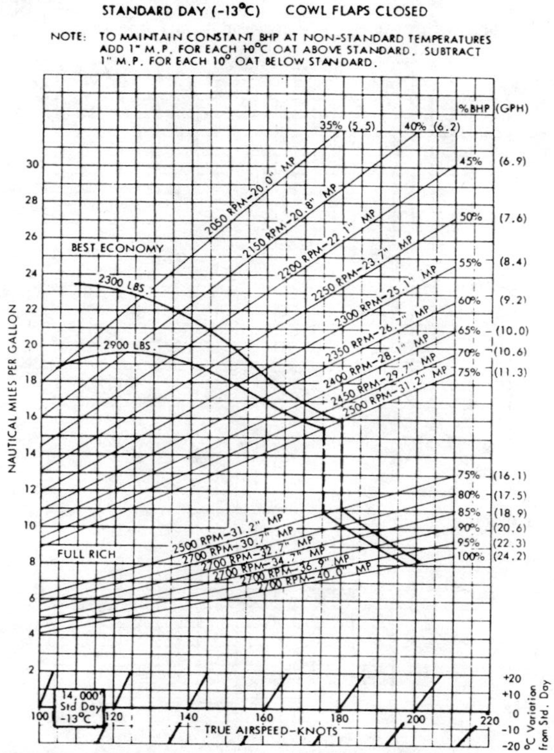

Fig. 10-64. Nautical miles per gallon at long-range cruise power at 14,000 feet.

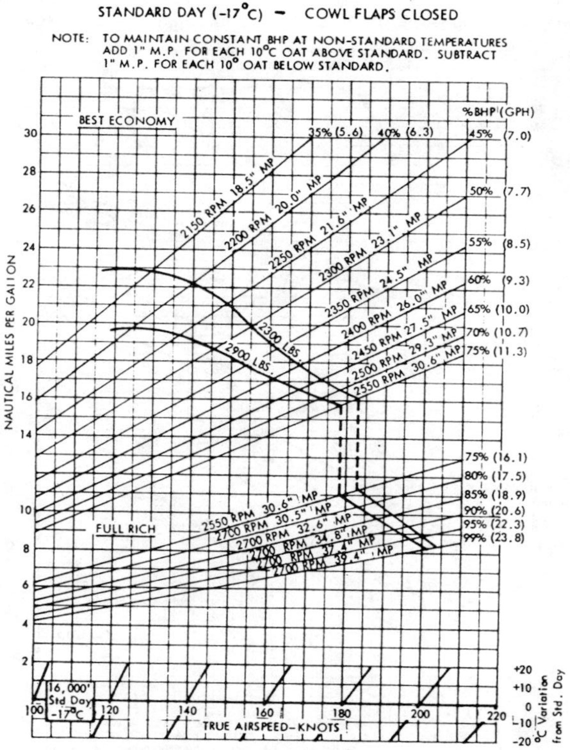

Fig. 10-65. Nautical miles per gallon at long-range cruise power at 16,000 feet.

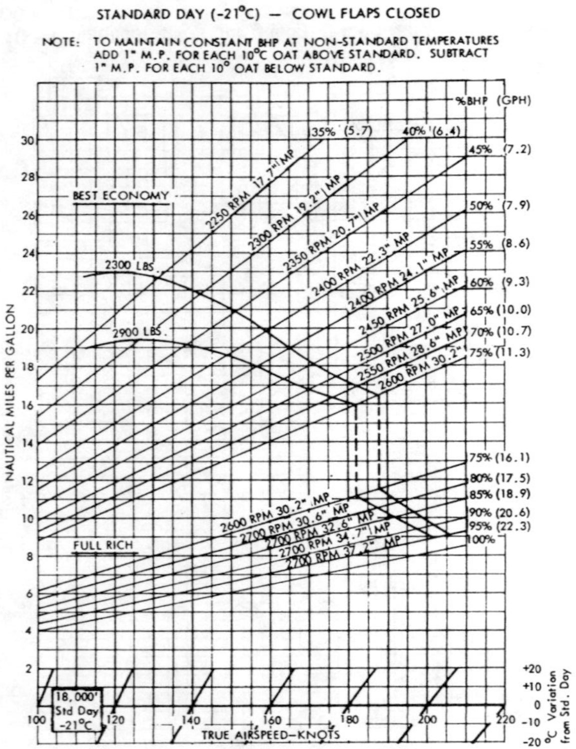

Fig. 10-66. Nautical miles per gallon at long-range cruise power at 18,000 feet.

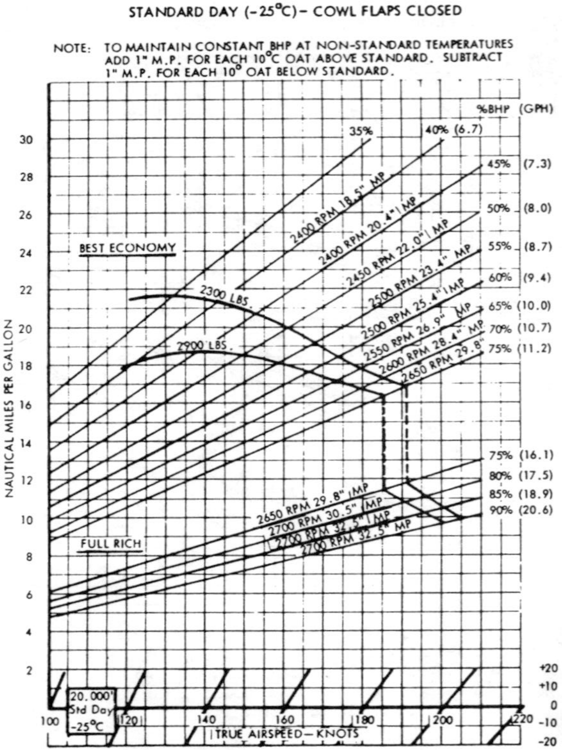

Fig. 10-67. Nautical miles per gallon at long-range cruise power at 20,000 feet.

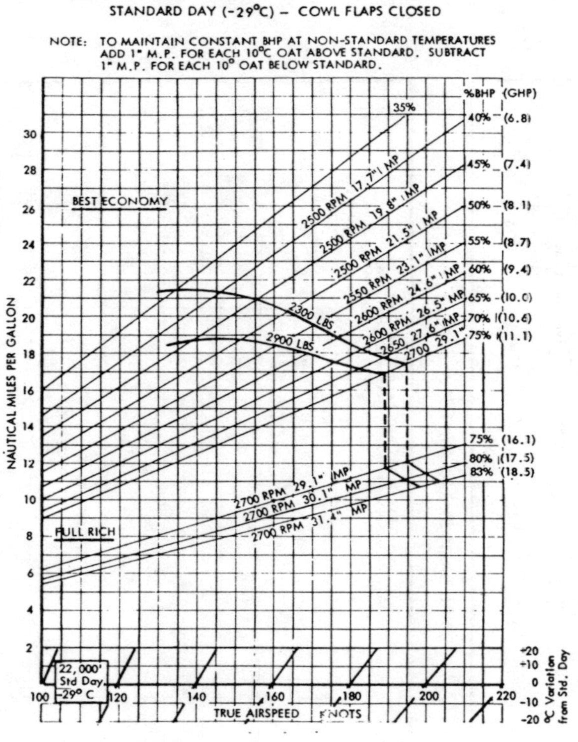

Fig. 10-68. Nautical miles per gallon at long-range cruise power at 22,000 feet.

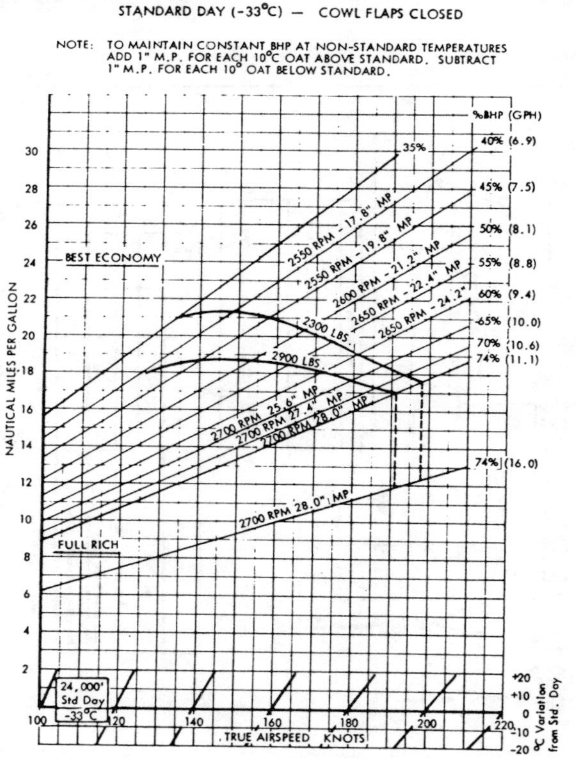

Fig. 10-69. Nautical miles per gallon at long-range cruise power at 24,000 feet.

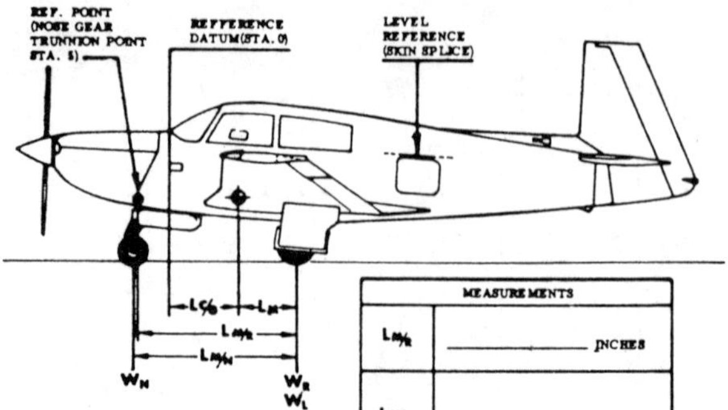

MEASUREMENTS	
$L_{M/R}$	______________ INCHES
$L_{M/N}$	______________ INCHES

SCALE POSITION AND SYMBOL	SCALE READING	TARE	NET WEIGHT
Nose Wheel (W_N)			
Right Main Wheel (W_R)			
Left Main Wheel (W_L)			
Basic Empty Weight, as Weighed (W_T)			

a. CG Forward of Main Wheels:

_____ LBS. Weight of Nose (W_N) X _____ IN. Distance Between Main and Nose Wheel Axle Centers ($L_{M/N}$) ÷ _____ LBS. Total Weight of Aircraft (W_T) = _____ IN. CG Forward of Main Wheels (L_M)

b. CG Aft of Datum (Station 0):

_____ IN. Distance from Center Nose Gear Trunion to Center of Main Wheel Axles (Horizontal) ($L_{M/R}$) − 5 IN. Distance from Nose Gear Trunion to Datum (Constant) − _____ IN. Result of Computation Above (L_M) = _____ IN. CG (FUS. STA.) Distance Aft of Datum. (Empty Weight CG) (L_{CG})

Fig. 10-70. Weighing procedure for the 231.

OWNERS WEIGHT AND BALANCE RECORD
(ENTER BELOW ALL WEIGHT CHANGE DATA FROM AIRCRAFT LOG BOOK)

AIRPLANE MODEL - M20 K	SERIAL NUMBER __________	FAA REGISTRATION NO. __________

DATE	DESCRIPTION OF MODIFICATION	WEIGHT CHANGE				RUNNING EMPTY WEIGHT			
		ADDED (+)		REMOVED (-)					
		Wt. (Pounds)	Arm (Inches)	Wt. (Pounds)	Arm (Inches)	Wt. (Pounds)	Moment /1000	Arm (Inches)	Useful Load
	BASIC EMPTY WEIGHT AS DELIVERED (W_T) (Includes Full Oil - 8 QTS)	—	—	—	—				

Fig. 10-71. Weight changes relative to weight and balance.

PROBLEM FORM

FAA REGISTRATION NO. _______________ M20K SERIAL NO. _______________

Step	ITEM	Sample Problem Pilot & Pass.		Your Problem	
		Weight (LBS)	Moment (LB-INS. 1000)	Weight (LBS)	Moment (LB-INS. /1000)
1	Aircraft Basic Empty Weight, (W_T) (From Page 6-6) Includes Full Oil ----8 QT. @ 1.875 LBS/QT (Sta - 22.19) (Sump assumed full for all flights)	1830	76.53		
2	Pilot Seat (*1)*	170.0	6.0 (2nd Pos.)		
	Copilot Seat (*2)*	170.0	5.78 (Fwd. Pos.)		
3	Left Rear Seat (*3)				
	Right Rear Seat (*4)				
4	Fuel (Max. Usable 72 Gal. 432 LBS @ sta 48.59)	360	17.49		
5	Baggage (Max. 120 LBS @ Sta 95.5)	110.0	10.51		
	Hat Rack (Max. 10 LBS @Sta 119.0)				
6	Loaded Aircraft Weight	2640	X		X
	Total Moment/1000	X	116.28	X	
7	**Refer to Page 6-8, Center-of Gravity Moment Envelope, to determine whether your aircraft loading is acceptable.**				

*Obtain the moment/1000 value for each seat position (FWD, MID, or AFT.) from loading computation graph below.

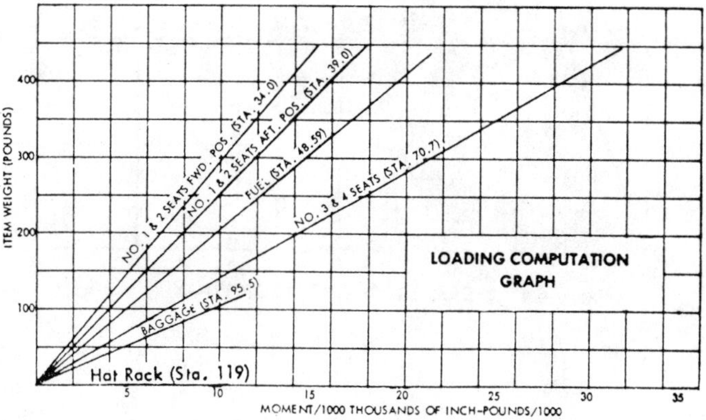

Fig. 10-72. Weight-and-balance problem form and loading computation graph.

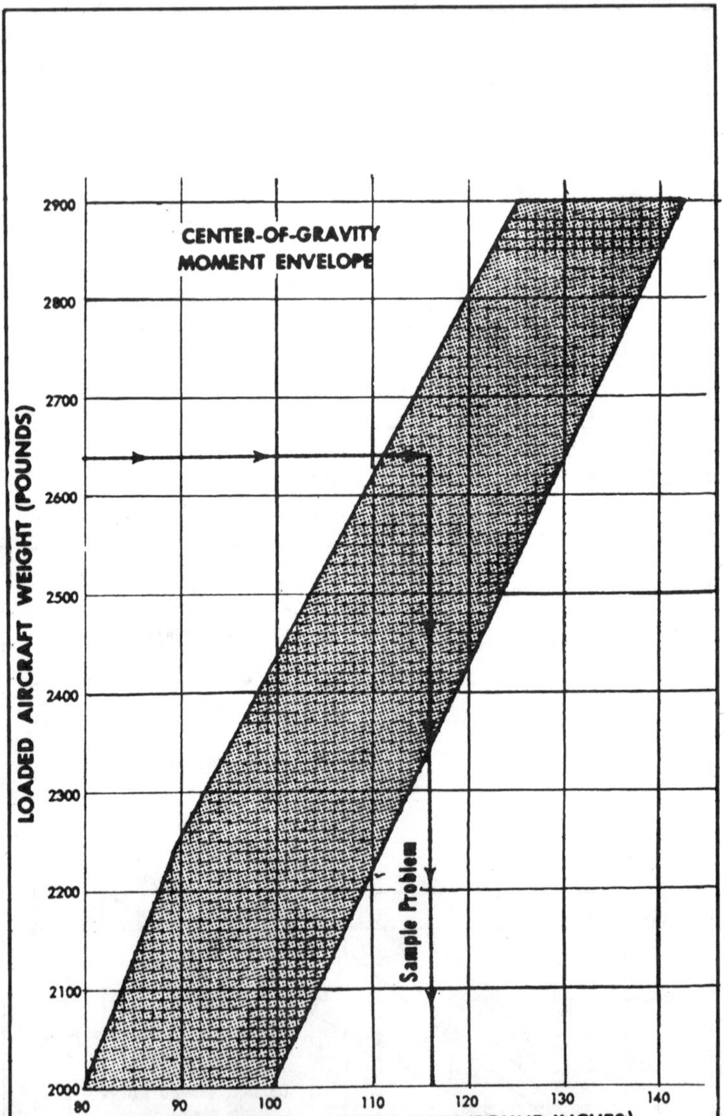

Fig. 10-73. Center-of-gravity moment envelope.

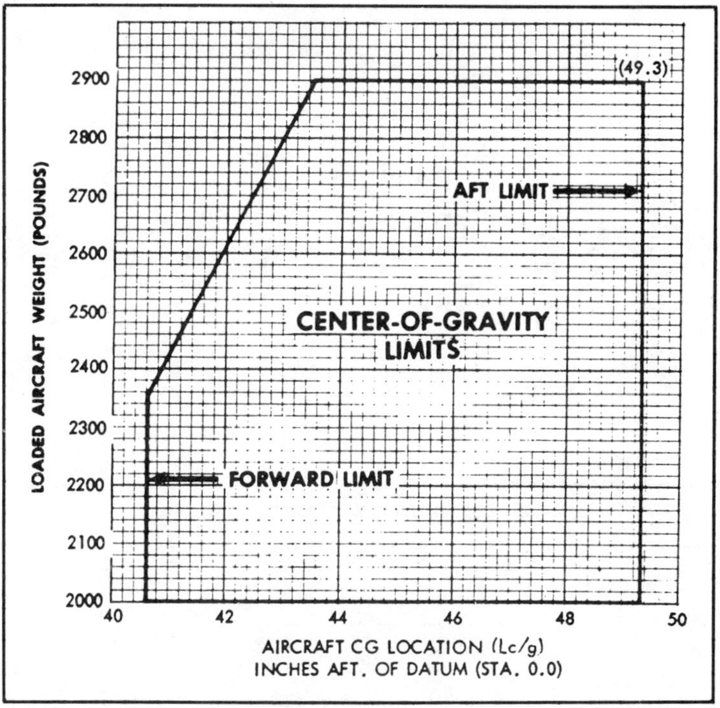

Fig. 10-74. Center-of-gravity limits.

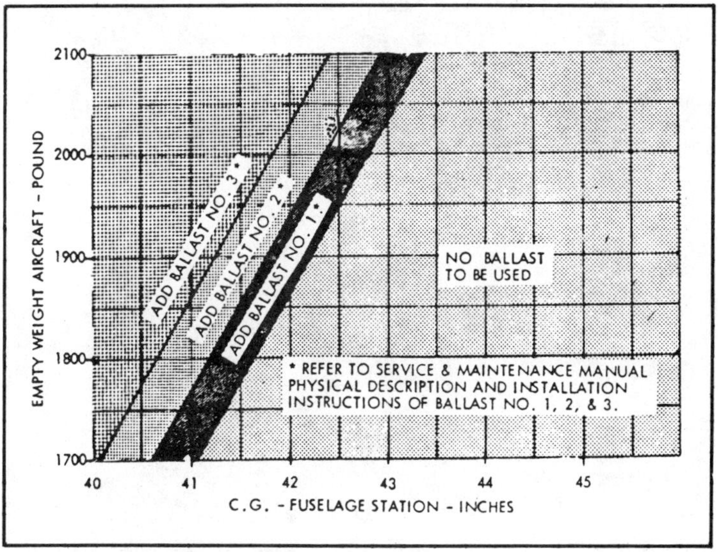

Fig. 10-75. Ballast weight chart.

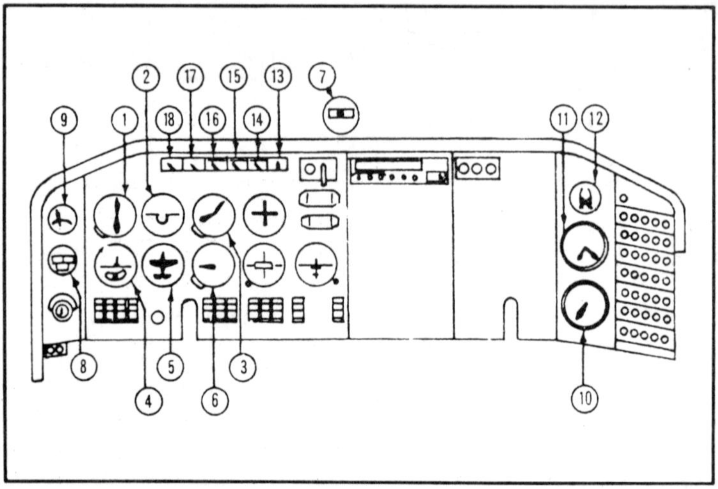

Fig. 10-76. Flight and panel instruments in the 231.

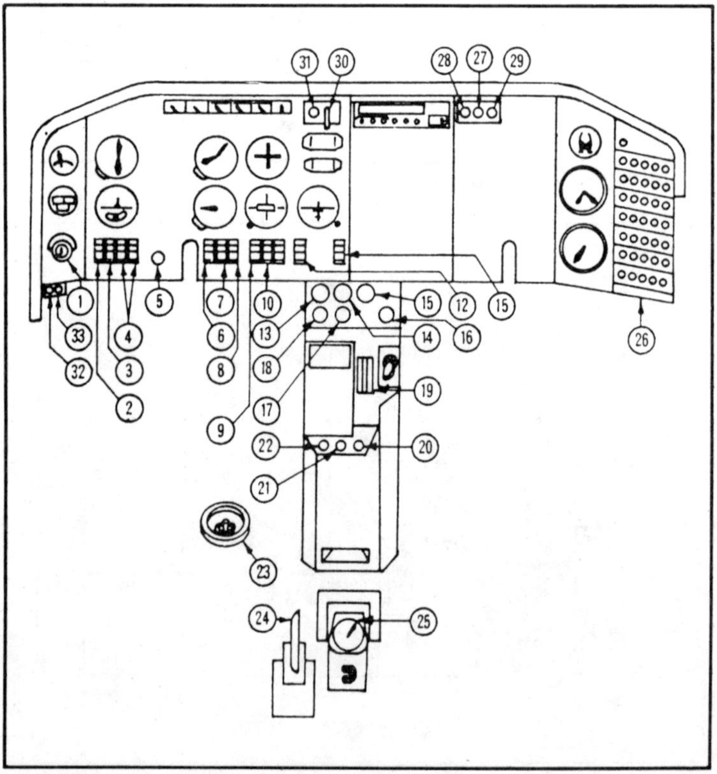

Fig. 10-77. Switches and controls in the 231.

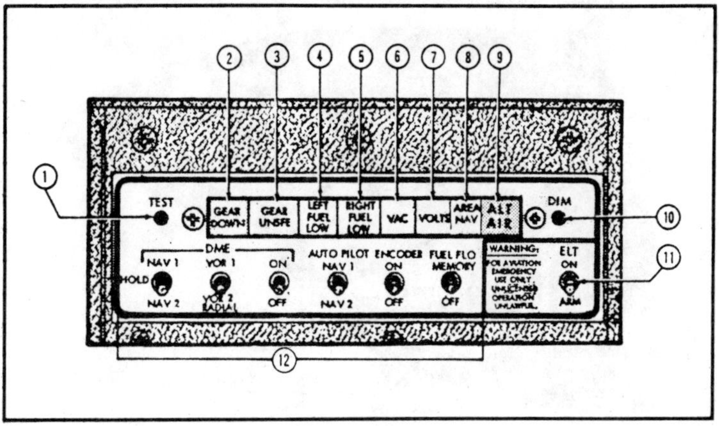

Fig. 10-78. Annunciator and switch panel in the 231.

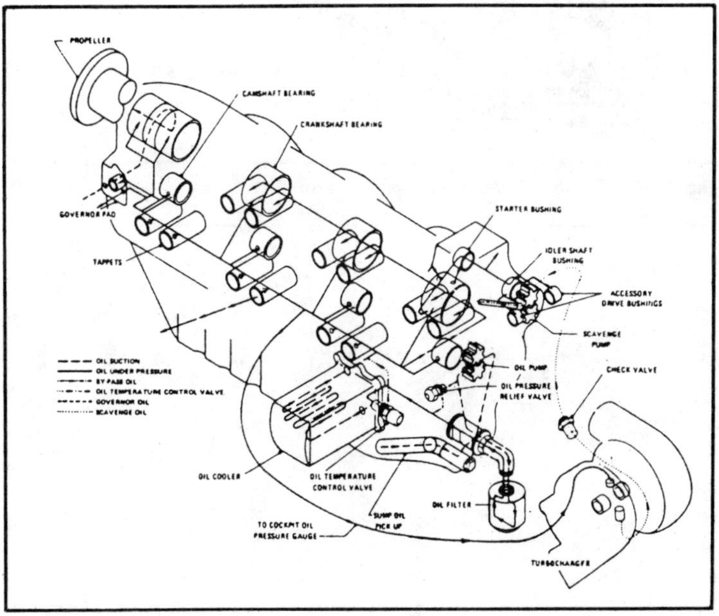

Fig. 10-79. The oil system in the 231.

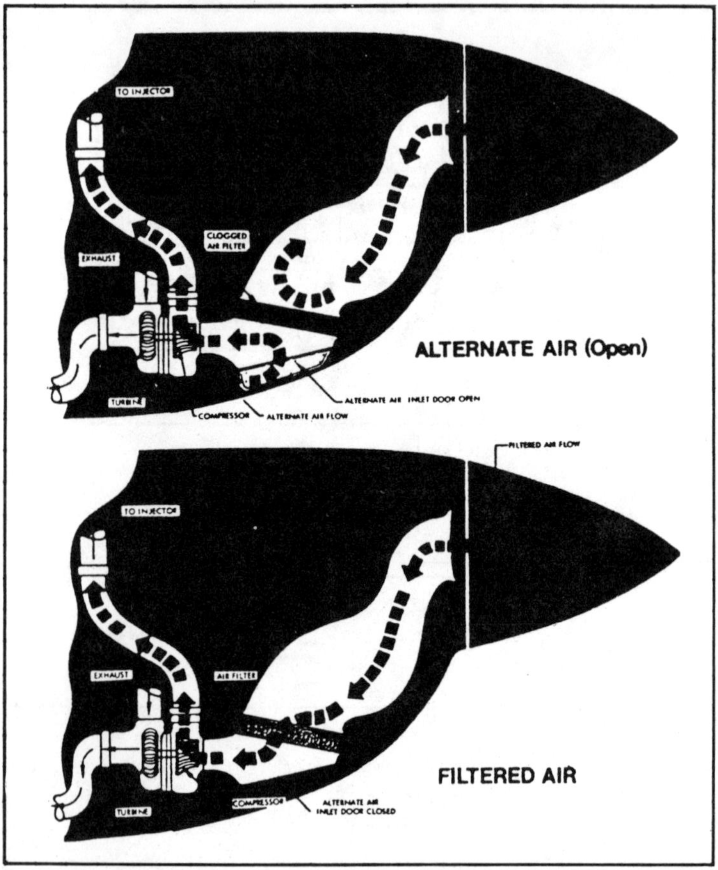

Fig. 10-80. Engine-air-induction system in the 231.

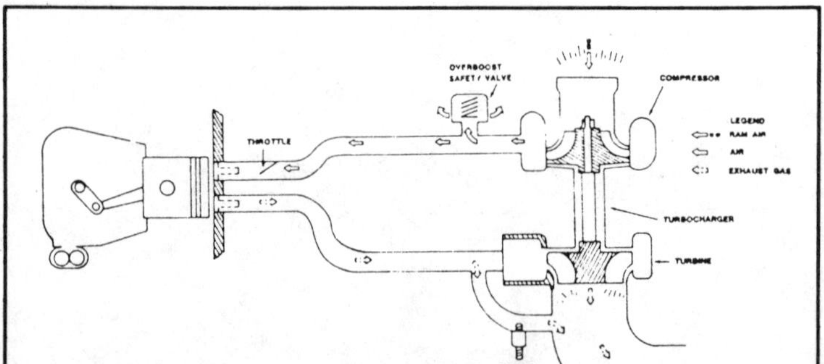

Fig. 10-81. Diagram of the turbocharger in the 231.

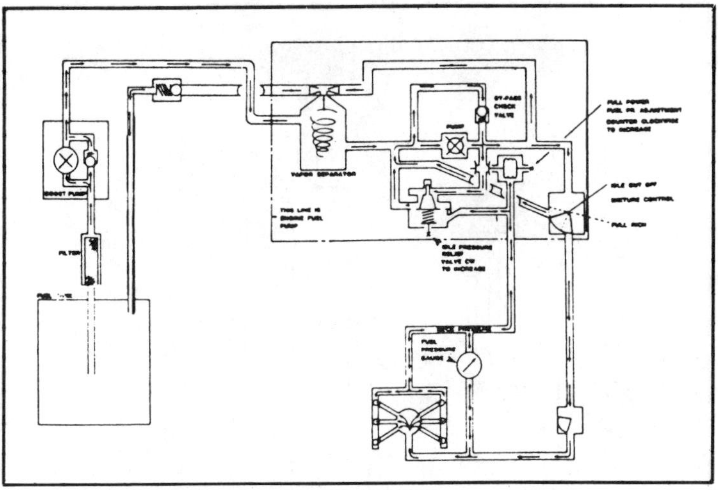

Fig. 10-82. The fuel-injection system in the 231.

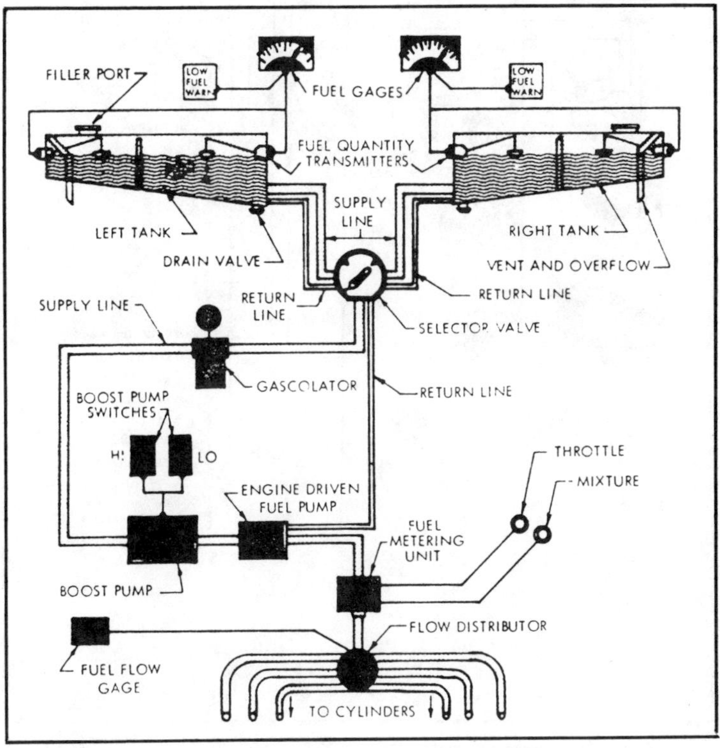

Fig. 10-83. The fuel system of the 231.

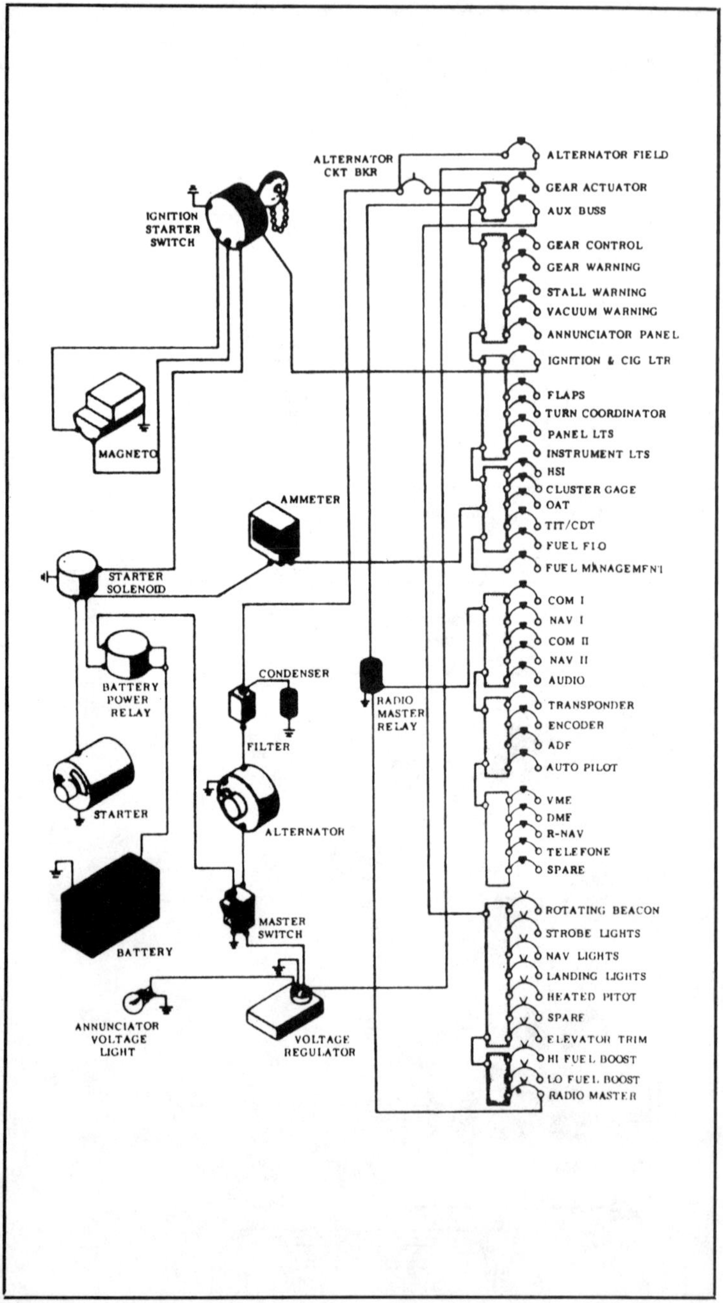

Fig. 10-84. The electrical system in the 231.

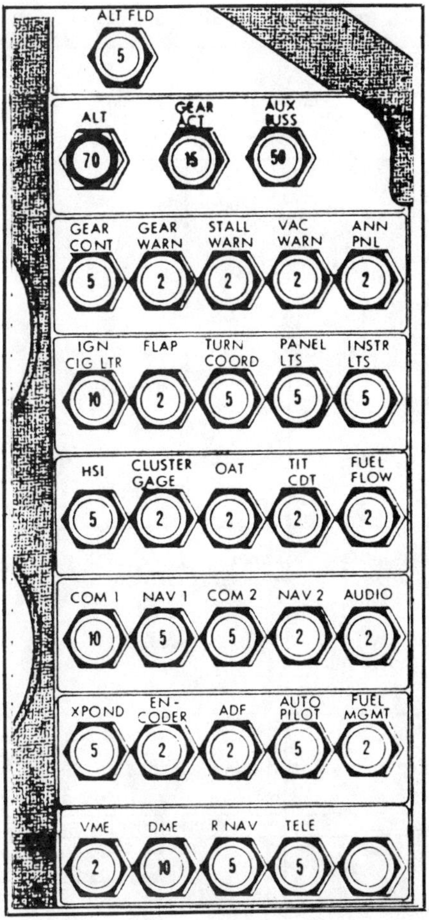

Fig. 10-85. The circuit-breaker panel in the 231.

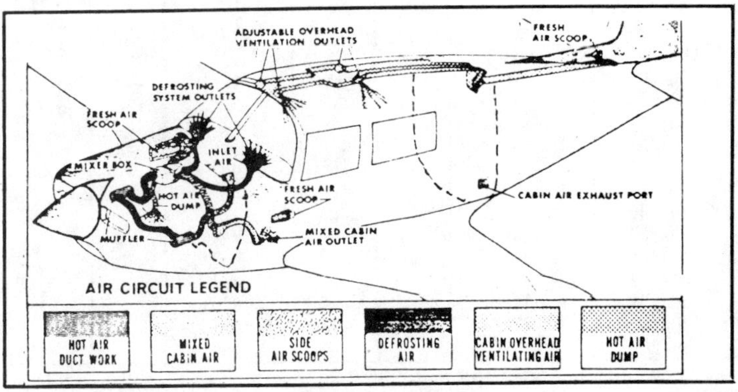

Fig. 10-86. The heating and ventilation system in the 231.

Table 10-1. Airspeed Calibrations Alternate Static Source.

IAS (MPH/KTS)	Gear & Flaps Up MPH/KTS	Gear & Flaps Down (15°) MPH/KTS	Gear & Flaps Down (33°) MPH/KTS
70/61	--	-2/-2	-4/-3
80/70	-2/-2	-3/-3	-6/-5
90/78	-4/-3	-5/-4	-8/-7
100/87	-4/-3	-7/-6	-9/-8
110/96	-5/-4	-8/-7	-11/-10
120/104	-6/-5	-8/-7	-11/-10
130/113	-6/-5	-8/-7	-12/-10
140/122	-7/-6	-- --	-- --
150/130	-7/-6	-- --	-- --
160/139	-7/-6	-- --	-- --
170/148	-7/-6	-- --	-- --
180/156	-7/-6	-- --	-- --
190/165	-4/-3	-- --	-- --
200/174	-4/-3	-- --	-- --
210/182	-5/-4	-- --	-- --
220/191	-5/-4	-- --	-- --
230/200	-6/-5	-- --	-- --

The minus sign indicates subtraction of the given numbers from IAS to obtain CAS assuming zero instrument error.

CONDITIONS: Storm Window and Vents: Closed
Defroster: Maximum

Power On

Table 10-2. Altimeter Corrections Alternate Static Source.

CONDITIONS: Storm Window and Vents: Closed, Defroster: Maximum, Power On

	SEA LEVEL			10,000 FT.		
IAS (MPH/KTS)	Gear & Flaps Up	Gear & Flaps Down 15°	Gear & Flaps Down 33°	Gear & Flaps Up	Gear & Flaps Down 15°	Gear & Flaps Down 33°
70/61	--	-10	-21	-4	-15	-28
80/70	-17	-20	-35	-21	-28	-39
90/78	-26	-37	-55	-36	-50	-76
100/87	-32	-54	-71	-43	-71	-99
110/96	-40	-55	-82	-55	-77	-102
120/104	-54	-63	-96	-73	-86	-130
130/113	-54	--	--	-84	--	--
140/122	-64	--	--	-87	--	--
150/130	-72	--	--	-99	--	--
160/139	-75	--	--	-101	--	--
170/148	-99	--	--	-134	--	--
180/156	-54	--	--	-73	--	--
190/165	-54	--	--	-73	--	--
200/174	-68	--	--	-94	--	--
210/182	-64	--	--	-83	--	--
220/191	-75	--	--	-103	--	--
230/200	-91	--	--	-125	--	--

The minus sign indicates subtraction of the given numbers from the indicated pressure altitude to obtain pressure altitude assuming zero instrument error.

Table 10-3. Stall Speeds.

ASSOCIATED CONDITIONS:

Gross Weight = 2740 LBS.
Forward CG
Power - Idle
Stall Speeds are indicated airspeeds in MPH & KNOTS and assume zero instrument error.

NOTE

Maximum altitude loss during stall recovery is approximately 290 feet

WEIGHT LBS.	ASSOCIATED CONDITIONS		Stall Speeds - MPH/KNOTS Angle of Bank 0°	20°	40°	60°
2740	Flaps & Gear Up	MPH	72	74	82	99
		KNOTS	63	64	71	86
	Flaps 15° Gear Down	MPH	66	68	76	95
		KNOTS	57	59	66	83
	Flaps 33° Gear Down	MPH	63	65	73	89
		KNOTS	55	57	63	77

EXAMPLE:

Given:	Weight	2740 LBS.
	Landing Gear	Down
	Flaps	33°
	Angle of Bank	20°
Find:	Stall Speed	57 Knots IAS 65 MPH IAS

Table 10-4. Maximum Performance Takeoff Distances.

ASSOCIATED CONDITIONS:

POWER------ FULL THROTTLE, 2700 RPM (Before Brake Release)
MIXTURE --- LEAN FOR SMOOTH OPERATION
FLAPS ------ 15°
LDG. GEAR-- EXTENDED UNTIL OBSTACLE CLEARED

RUNWAY------------- PAVED, LEVEL, DRY SURFACE
WEIGHT------------- 2740 LBS.
TAKEOFF SPEED----- 71 MPH/62 KTS IAS
CLIMB OUT----------- 76 MPH/66 KTS IAS
COWL FLAPS--------- FULL OPEN

		PRESSURE ALTITUDE									
		Sea Level		2000 FT.		4000 FT.		6000 FT.		8000 FT.	
Wind Component Down Runway Knots	OAT °C	Ground Roll Feet	Total Over 50 Ft. Obstacle Feet	Ground Roll Feet	Total Over 50 Ft. Obstacle Feet	Ground Roll Feet	Total Over 50 Ft. Obstacle Feet	Ground Roll Feet	Total Over 50 Ft. Obstacle Feet	Ground Roll Feet	Total Over 50 Ft. Obstacle Feet
0	-20	679	1179	803	1387	1029	1740	1330	2175	1692	2678
	-10	732	1267	873	1498	1118	1879	1446	2351	1843	2898
	0	793	1364	946	1613	1211	2024	1566	2532	1993	3119
	10	857	1465	1022	1732	1309	2175	1692	2720	2153	3352
	20	924	1570	1101	1856	1410	2330	1823	2916	2329	3603
	30	993	1678	1183	1983	1516	2491	1960	3119	----	----
	40	1064	1789	1269	2116	1625	2657	2101	3327	----	----
10	-20	608	1071	722	1264	929	1591	1205	1994	1538	2461
	-10	657	1153	787	1368	1016	1726	1312	2158	1678	2667
	0	713	1243	854	1475	1097	1856	1424	2328	1818	2875
	10	772	1337	924	1586	1188	1998	1541	2505	1967	3094
	20	834	1436	997	1702	1282	2143	1663	2689	2131	3329
	30	898	1537	1073	1821	1380	2294	1790	2879	----	----
	40	963	1640	1153	1946	1482	2450	1922	3075	----	----
20	-20	548	974	653	1154	843	1456	1097	1830	1405	2265
	-10	593	1050	713	1251	919	1577	1197	1984	1536	2459
	0	645	1135	775	1350	999	1704	1300	2143	1666	2654
	10	700	1223	840	1454	1083	1836	1409	2308	1805	2859
	20	757	1314	908	1563	1170	1972	1523	1481	1958	3081
	30	816	1408	978	1674	1262	2114	1642	2661	----	----
	40	877	1506	1053	1791	1357	2261	1765	2845	----	----

NOTE: 1) Maximum demonstrated crosswind velocity is 12 MPH (11 Knots). 2) Where distance value has been deleted, climb performance after lift off is less than 150 ft./min. 3) Conditions of high humidity can result in an increase of up to 10% to the above take-off distances.

Table 10-5. Normal Performance Takeoff Distances.

ASSOCIATED CONDITIONS:

POWER------ FULL THROTTLE, 2700 RPM (Before Brake Release)
MIXTURE --- LEAN FOR SMOOTH OPERATION
FLAPS ------ 15°
LDG. GEAR-- EXTENDED UNTIL OBSTACLE CLEARED

RUNWAY-------------- PAVED, LEVEL, DRY SURFACE
WEIGHT--------------- 2740 LBS.
TAKEOFF SPEED------73 MPH/63 KTS IAS
CLIMB OUT----------- 82 MPH/71 KTS IAS
COWL FLAPS---------- FULL OPEN

		PRESSURE ALTITUDE									
		Sea Level		2000 FT.		4000 FT.		6000 FT.		8000 FT.	
Wind Component Down Runway Knots	OAT °C	Ground Roll Feet	Total Over 50 Ft. Obstacle Feet	Ground Roll Feet	Total Over 50 Ft. Obstacle Feet	Ground Roll Feet	Total Over 50 Ft. Obstacle Feet	Ground Roll Feet	Total Over 50 Ft. Obstacle Feet	Ground Roll Feet	Total Over 50 Ft. Obstacle Feet
0	-20	704	1374	854	1646	1049	2074	1392	2808	1778	3820
	-10	765	1482	928	1776	1140	2237	1513	3028	1933	4118
	0	829	1594	1005	1910	1235	2406	1639	3256	2094	4426
	10	896	1711	1086	2050	1334	2581	1771	3494	2262	4746
	20	965	1831	1170	2194	1438	2764	1908	3738	2437	5077
	30	1037	1955	1258	2344	1545	2951	2051	3992	----	----
	40	1112	2084	1349	2498	1657	3145	2199	4253	----	----
10	-20	632	1255	769	1507	948	1906	1263	2490	1619	3537
	-10	688	1356	837	1629	1032	2059	1375	2797	1763	3818
	0	747	1460	908	1754	1120	2217	1492	3011	1913	4109
	10	814	1575	983	1885	1212	2382	1615	3236	2070	4412
	20	872	1681	1061	2021	1309	2555	1743	3466	2233	4725
	30	939	1798	1143	2162	1408	2730	1876	3705	----	----
	40	1008	1919	1227	2306	1513	2914	2014	3952	----	----
20	-20	570	1446	696	1381	862	1753	1151	2389	1480	3275
	-10	622	1240	760	1495	940	1897	1255	2583	1615	3541
	0	676	1338	826	1613	1021	2045	1365	2786	1755	3815
	10	738	1445	895	1736	1107	2200	1479	2997	1901	4101
	20	793	1546	967	1862	1197	2362	1598	3214	2054	4397
	30	854	1654	1043	1995	1290	2528	1723	3441	----	----
	40	919	1768	1122	2131	1387	2700	1852	3674	----	----

NOTE: 1) Maximum demonstrated crosswind velocity is 12 MPH (11 Knots). 2) Where distance value has been deleted, climb performance after lift off is less than 150 ft./min. 3) Conditions of high humidity can result in an increase of up to 10% to the above take-off distances.

Table 10-6. Cruise Performance at Economy Cruise at Sea Level.

MIXTURE SETTING:
Lean mixture in accordance with instructions in Section IV.

RPM	MAN PRES (IN. HG)	% BHP	FUEL (GAL/HR)	FUEL (LBS/HR)	TRUE AIRSPEED MPH/KNOTS 2740 LBS	TRUE AIRSPEED MPH/KNOTS 2300 LBS	ENDURANCE (HR: MIN)	RANGE (NAUT MI) 2740 LBS	RANGE (NAUT MI) 2300 LBS
2700	23.5	75	10.8	65.0	180/156	182/158	5:00	780	790
	22.0	70	10.3	61.5	175/152	178/155	5:20	810	826
	21.0	65	9.7	58.0	169/147	172/149	5:42	843	860
	19.5	60	9.2	55.0	163/142	167/145	6:08	864	891
	18.0	55	8.6	51.5	157/136	161/140	6:32	886	917
	14.0	40	7.0	42.0	130/113	138/120	8:20	921	977
2600	24.5	75	10.5	63.0	180/156	182/158	5:12	812	808
	23.0	70	10.0	60.0	175/152	178/155	5:30	843	856
	20.5	60	8.9	53.5	163/142	167/145	6:20	895	921
	19.0	55	8.3	50.0	157/136	161/140	6:45	917	951
	17.5	50	7.8	47.0	150/130	154/134	7:20	938	982
	15.0	40	6.8	40.5	130/113	138/120	8:35	951	1025
2400	27.0	75	10.3	62.5	180/156	182/158	5:20	834	827
	24.0	65	9.2	55.0	169/147	172/149	6:10	895	918
	21.0	55	8.1	48.5	157/136	161/140	7:00	951	977
	17.5	45	7.0	42.0	140/122	147/128	8:10	990	1036
	15.5	39	6.4	38.5	128/111	137/119	9:05	999	1064
2200	27.0	68	9.3	55.5	173/150	176/153	6:00	900	918
	22.5	55	7.8	47.0	157/136	161/140	7:15	986	1015
	21.0	50	7.3	44.0	150/130	154/134	7:30	1019	1050
	19.0	45	6.8	40.5	140/122	147/128	8:10	1079	1082
	17.5	37	5.9	35.5	122/106	133/116	9:25	1110	1125
2000	24.0	53	7.4	44.5	154/134	158/137	7:40	1030	1064
	23.0	50	7.1	42.5	150/130	154/134	8:10	1053	1090
	21.0	45	6.5	39.0	140/122	147/128	8:52	1079	1138
	17.0	36	5.6	33.5	119/103	130/113	10:40	1110	1221

Table 10-7. Cruise Performance at Economy Cruise at 2,000 Feet.

MIXTURE SETTING:
Lean mixture in accordance with instructions in Section IV.

					TRUE AIRSPEED MPH/KNOTS		ENDURANCE	RANGE (NAUT MI)	
RPM	MAN PRES (IN. HG)	% BHP	FUEL (GAL/HR)	FUEL (LBS/HR)	2740 LBS	2300 LBS	(HR: MIN)	2740 LBS	2300 LBS
2700	23.3	75	10.8	64.6	183/159	186/161	5:00	800	808
	20.6	65	9.7	58.0	171/149	176/153	5:40	849	869
	18.0	55	8.6	51.5	159/138	164/142	6:30	897	923
	14.8	43	7.3	43.5	139/121	146/127	7:50	947	994
2600	24.4	75	10.5	63.0	183/159	186/161	5:10	821	831
	21.6	65	9.4	56.5	172/149	176/153	5:50	869	892
	18.8	55	8.3	50.0	159/138	164/142	6:50	943	970
	15.2	42	6.9	41.5	136/118	145/126	8:20	982	1049
2400	26.8	75	10.3	61.4	183/159	186/161	5:20	847	858
	23.6	65	9.2	55.0	172/149	176/153	6:05	906	930
	20.4	55	8.1	48.5	159/138	164/142	7:00	966	994
	16.0	41	6.6	39.5	134/116	144/125	8:40	1004	1082
2200	25.4	64	8.7	52.4	171/149	175/152	6:20	943	962
	22.2	55	7.8	47.0	159/138	164/142	7:15	1000	1029
	18.7	45	6.8	40.5	142/123	150/130	8:30	1045	1105
	16.5	39	6.1	36.5	129/112	140/122	9:30	1064	1159
2000	22.5	50	7.1	42.5	151/131	157/136	8:05	1058	1099
	20.5	45	6.5	39.0	142/123	150/130	8:50	1086	1148
	17.4	37	5.7	34.0	124/108	136/118	10:20	1115	1218

Table 10-8. Cruise Performance at Economy Cruise at 4,000 Feet.

MIXTURE SETTING:
Lean mixture in accordance
with instructions in Section IV.

					TRUE AIRSPEED MPH/KNOTS		ENDURANCE	RANGE (NAUT MI)	
RPM	MAN PRES (IN. HG)	% BHP	FUEL (GAL/HR)	FUEL (LBS/HR)	2740 LBS	2300 LBS	(HR: MIN)	2740 LBS	2300 LBS
2700	23.2	75	10.8	64.6	186/161	189/164	5:00	805	820
	20.5	65	9.7	58.0	175/152	179/155	5:40	861	878
	17.9	55	8.6	51.5	161/140	166/144	6:30	910	936
	14.8	43	7.3	43.5	140/121	148/128	7:45	937	992
2600	24.4	75	10.5	63.0	186/161	189/164	5:10	831	847
	21.5	65	9.4	56.5	175/152	179/155	5:52	891	909
	18.7	55	8.3	50.0	161/140	166/144	6:43	940	967
	15.0	42	6.9	41.5	138/120	146/127	8:30	1020	1079
2400	26.2	74	10.2	60.8	185/161	188/163	5:12	837	847
	23.3	65	9.2	55.0	175/152	179/155	5:57	904	922
	20.2	55	8.1	48.5	161/140	166/144	6:57	973	1000
	15.8	41	6.6	39.5	135/117	145/126	8:40	1013	1091
2200	24.4	62	8.5	51.5	171/148	175/152	6:25	949	975
	22.0	55	7.8	47.0	161/140	166/144	7:10	1003	1032
	18.6	45	6.8	40.5	144/125	151/131	8:28	1058	1109
	16.4	39	6.1	36.5	130/113	140/121	9:28	1069	1145
2000	22.4	50	7.1	42.5	154/134	159/138	8:00	1072	1104
	20.4	45	6.5	39.0	144/125	151/131	8:45	1093	1146
	17.2	37	5.7	34.0	124/108	137/119	10:14	1105	1217

Table 10-9. Cruise Performance at Economy Cruise at 6,000 feet.

MIXTURE SETTING:
Lean mixture in accordance
with instructions in Section IV.

RPM	MAN PRES (IN. HG)	% BHP	FUEL (GAL/HR)	FUEL (LBS/HR)	TRUE AIRSPEED MPH/KNOTS 2740 LBS	TRUE AIRSPEED MPH/KNOTS 2300 LBS	ENDURANCE (HR: MIN)	RANGE (NAUT MI) 2740 LBS	RANGE (NAUT MI) 2300 LBS
2700	23.1	75	10.8	64.7	189/164	192/167	5:00	820	835
	20.4	65	9.7	58.0	178/154	181/157	5:40	871	888
	17.8	55	8.6	51.5	164/142	170/148	6:25	911	949
	15.2	45	7.5	45.0	145/126	155/134	7:28	940	1000
2600	24.1	75	10.5	63.0	189/164	192/167	5:10	847	862
	21.3	65	9.4	56.5	178/154	181/157	5:50	898	915
	18.5	55	8.3	50.1	164/142	170/148	6:38	941	981
	15.4	44	7.2	43.0	144/125	152/13[illegible]	7:50	979	1034
2400	24.4	70	9.7	58.0	184/160	187/162	5:40	906	918
	22.8	65	9.2	55.0	178/154	181/157	6:00	924	942
	19.8	55	8.1	48.6	164/142	170/148	6:50	970	1011
	16.2	43	6.8	40.6	141/122	151/131	8:15	1006	1080
2200	23.6	60	8.3	50.0	171/148	176/153	6:38	981	1014
	21.8	55	7.8	47.1	164/142	170/148	7:10	1017	1060
	20.0	50	7.3	44.1	155/135	162/141	7:40	1035	1081
	17.2	42	6.4	38.5	139/121	150/130	8:50	1068	1148
2000	21.3	47	6.7	40.2	150/130	159/138	8:20	1083	1150
	18.8	41	6.1	36.4	136/118	147/128	9:20	1101	1194

Table 10-10. Cruise Performance at Economy Cruise at 8,000 Feet.

MIXTURE SETTING:
Lean mixture in accordance with instructions in Section IV.

RPM	MAN PRES (IN. HG)	% BHP	FUEL (GAL/HR)	FUEL (LBS/HR)	TRUE AIRSPEED MPH/KNOTS 2740 LBS	TRUE AIRSPEED MPH/KNOTS 2300 LBS	ENDURANCE (HR:MIN)	RANGE (NAUT MI) 2740 LBS	RANGE (NAUT MI) 2300 LBS
2700	23.6	75	10.8	64.7	195/169	197/171	5:00	835	850
	21.7	70	10.3	61.6	187/162	191/166	5:15	850	871
	20.4	65	9.7	58.0	181/157	185/161	5:37	881	904
	19.0	60	9.2	55.1	174/151	179/156	6:00	906	936
	17.8	55	8.6	51.5	167/145	173/150	6:27	928	961
	14.8	44	7.4	44.4	146/127	155/135	7:31	954	1014
2600	23.0	71	10.1	60.6	189/164	193/168	5:20	869	890
	21.2	65	9.4	56.6	181/157	185/161	5:48	910	933
	19.8	60	8.8	53.3	174/151	179/156	6:10	931	962
	18.6	55	8.3	50.1	167/145	173/150	6:37	952	985
	17.0	50	7.8	46.8	158/137	165/143	7:05	970	1012
	15.2	43	7.6	45.5	144/125	153/133	7:55	989	1053
2400	22.8	64	9.1	54.4	181/157	185/161	6:04	946	970
	21.3	60	8.6	51.6	174/151	179/156	6:21	958	990
	19.8	55	8.1	48.6	167/145	173/150	6:50	984	1018
	18.2	50	7.5	45.5	158/137	165/143	7:20	1004	1048
	15.5	42	6.7	40.0	141/122	151/131	8:20	1016	1091
2200	22.0	55	7.8	47.1	167/145	173/150	7:05	1020	1055
	20.0	50	7.3	44.1	158/137	165/143	7:40	1050	1096
	16.8	41	6.3	38.0	138/120	150/130	8:54	1068	1157
2000	20.3	45	6.5	39.0	148/129	156/136	8:35	1102	1167
	18.2	40	6.0	35.7	135/117	147/128	9:28	1110	1211

Table 10-11. Cruise Performance at Economy Cruise at 10,000 Feet.

MIXTURE SETTING:
Lean mixture in accordance with instructions in Section IV.

RPM	MAN PRES (IN. HG)	% BHP	FUEL (GAL/HR)	FUEL (LBS/HR)	TRUE AIRSPEED MPH/KNOTS 2740 LBS	TRUE AIRSPEED MPH/KNOTS 2300 LBS	ENDURANCE (HR: MIN)	RANGE (NAUT MI) 2740 LBS	RANGE (NAUT MI) 2300 LBS
2700	21.4	70	10.3	61.5	190/165	195/169	5:15	851	887
	20.2	65	9.7	58.0	184/160	189/164	5:31	877	904
	18.8	60	9.2	55.1	178/155	183/159	5:55	907	940
	17.6	55	8.6	51.5	170/148	176/153	6:20	927	969
	16.3	50	7.7	46.2	161/140	169/147	6:50	944	1004
	15.0	45	7.5	45.0	150/130	160/139	7:25	949	1030
2600	21.1	65	9.4	56.5	184/160	189/164	5:48	910	951
	19.8	60	8.9	53.3	178/155	183/159	6:10	938	980
	18.3	55	8.4	50.1	170/148	176/153	6:35	962	1007
	17.0	50	7.8	46.7	161/140	169/147	7:03	980	1036
	15.3	44	7.2	43.0	149/129	159/138	7:45	988	1069
2400	21.0	60	8.6	51.7	178/155	183/159	6:20	961	1006
	19.5	55	8.1	48.5	170/148	176/153	6:47	908	1037
	18.0	50	7.6	45.5	161/140	169/147	7:18	1076	1073
	16.2	44	7.1	42.5	149/129	159/138	8:01	1025	1106
2200	21.0	51	7.4	44.5	164/142	170/148	7:28	1051	1105
	17.8	44	6.7	39.9	149/129	159/138	8:24	1068	1159

Table 10-12. Cruise Performance at Economy Cruise at 12,000 Feet.

MIXTURE SETTING:
Lean mixture in accordance with instructions in Section IV.

RPM	MAN PRES (IN. HG)	% BHP	FUEL (GAL/HR)	FUEL (LBS/HR)	TRUE AIRSPEED MPH/KNOTS 2740 LBS	TRUE AIRSPEED MPH/KNOTS 2300 LBS	ENDURANCE (HR: MIN)	RANGE (NAUT MI) 2740 LBS	RANGE (NAUT MI) 2300 LBS
2700	19.8	64	9.6	57.5	185/161	191/166	5:40	869	882
	18.8	60	9.2	55.1	180/156	186/162	5:55	898	947
	17.5	55	8.6	51.5	173/150	179/150	6:20	923	977
	16.2	50	7.7	46.2	165/143	171/149	6:45	947	1005
	14.6	44	7.4	44.4	150/130	160/139	7:28	955	1037
2600	19.6	60	8.9	53.3	180/156	186/162	6:10	896	973
	18.2	55	8.4	50.1	173/150	179/156	6:34	959	1011
	16.8	50	7.8	46.7	165/143	171/149	7:00	983	1043
	15.5	45	7.3	43.5	154/134	162/141	7:35	990	1069
	14.8	43	7.1	42.5	149/129	159/138	7:50	990	1081
2400	19.6	56	8.2	49.3	174/151	180/156	6:40	973	1039
	17.9	50	7.6	45.5	165/143	171/149	7:20	1016	1092
	16.4	45	7.0	42.0	154/134	162/141	7:54	1030	1113
	15.2	42	6.7	40.0	145/126	157/136	8:20	1025	1133
2200	19.4	49	7.2	43.0	162/141	170/147	7:40	1059	1126
	18.0	45	6.8	40.5	154/134	162/141	8:10	1080	1151
	17.8	42	6.4	38.5	145/126	157/136	8:35	1081	1180

Table 10-13. Cruise Performance at Economy Cruise at 14,000 Feet.

MIXTURE SETTING:
Lean mixture in accordance
with instructions in Section IV.

					TRUE AIRSPEED MPH/KNOTS		ENDURANCE	RANGE (NAUT MI)	
RPM	MAN PRES (IN. HG)	% BHP	FUEL (GAL/HR)	FUEL (LBS/HR)	2740 LBS	2300 LBS	(HR: MIN)	2740 LBS	2300 LBS
2700	18.4	58	8.9	53.5	180/156	186/162	6:05	912	964
	17.4	55	8.6	51.5	175/152	182/158	6:17	929	986
	16.1	50	7.7	46.2	165/143	175/152	6:42	946	1012
	14.4	44	7.4	44.4	150/130	164/142	7:20	948	1043
2600	18.2	56	8.4	50.6	176/153	184/160	6:30	994	1040
	16.7	50	7.8	46.7	165/143	175/152	7:00	1001	1064
	14.8	43	7.6	45.5	149/129	161/122	7:48	1006	951
2400	18.2	52	7.8	46.6	169/147	177/154	7:05	1000	1068
	16.2	45	7.0	42.0	154/134	165/143	7:45	1016	1116
	15.3	42	6.7	40.0	145/126	160/139	8:10	1016	1138
2200									
2000									

Table 10-14. Cruise Performance Using Best Power at Sea Level.

MIXTURE SETTING:
1. Use FULL RICH mixture above 75% power. 2. Lean mixture in accordance with instructions in Section IV at 75% power and below.

RPM	MAN PRES (IN. HG)	% BHP	FUEL (GAL/HR)	FUEL (LBS/HR)	TRUE AIRSPEED MPH/KNOTS 2740 LBS	TRUE AIRSPEED MPH/KNOTS 2300 LBS	ENDURANCE (HR:MIN)	RANGE (NAUT MI) 2740 LBS	RANGE (NAUT MI) 2300 LBS
2700	30.2	100	18.4	110.5	201/175	203/176	2:35	452	454
	26.2	85	15.8	94.5	188/163	191/166	3:15	530	539
	23.5	75	12.5	75.0	180/156	182/158	4:12	655	663
	21.0	65	11.3	67.5	169/147	172/149	4:45	698	707
	18.0	55	10.0	60.0	157/136	161/140	5:30	748	770
	14.9	43	8.7	52.0	136/118	143/124	6:30	767	806
2600	30.1	94	14.8	89.0	194/169	198/172	2:55	492	501
	27.5	85	13.7	82.0	188/163	191/166	3:15	529	539
	24.7	75	12.3	74.0	180/156	182/158	4:20	675	684
	21.8	65	10.9	65.5	169/147	172/149	4:55	722	732
	19.0	55	9.8	59.0	157/136	161/140	5:45	782	805
	15.1	41	8.2	49.0	132/115	140/122	6:55	795	843
2400	28.7	80	14.8	88.5	184/160	187/162	3:25	546	553
	27.0	75	12.0	72.0	180/156	182/158	4:30	702	711
	24.0	65	10.7	64.0	169/147	172/149	5:05	747	757
	21.0	55	9.5	57.0	157/136	161/140	5:50	793	816
	16.2	41	7.5	45.2	127/110	137/119	7:30	825	892
2200	27.1	68	11.3	67.5	173/150	176/153	5:05	762	777
	22.5	55	9.2	55.0	157/136	161/140	6:05	827	851
	20.8	50	8.6	51.5	150/130	154/134	6:30	845	871
	19.0	45	7.8	47.0	140/122	147/128	7:05	864	906
	17.2	40	7.5	44.8	130/113	138/120	7:40	865	919
2000	24.0	53	8.6	51.5	154/134	158/137	6:30	871	890
	23.0	50	8.3	50.0	150/130	154/134	6:50	888	915
	21.0	45	7.8	46.5	140/122	147/128	7:20	894	938
	17.0	36	6.7	40.0	119/103	130/113	8:33	880	966

Table 10-15. Cruise Performance Using Best Power at 2,000 Feet.

MIXTURE SETTING:
1. Use FULL RICH mixture above 75% power. 2. Lean mixture in accordance with instructions in Section IV at 75% power and below.

RPM	MAN PRES (IN. HG)	% BHP	FUEL (GAL/HR)	FUEL (LBS/HR)	TRUE AIRSPEED MPH/KNOTS 2740 LBS	TRUE AIRSPEED MPH/KNOTS 2300 LBS	ENDURANCE (HR: MIN)	RANGE (NAUT MI) 2740 LBS	RANGE (NAUT MI) 2300 LBS
2700	28.1	93	14.9	89.5	199/173	201/175	3:00	500	500
	26.0	85	13.8	83.0	192/167	195/169	3:10	534	543
	23.3	75	12.6	75.4	183/159	186/161	4:15	669	682
	20.6	65	11.3	67.5	171/149	176/153	4:50	714	730
	18.0	55	10.0	60.0	159/138	164/142	5:25	754	778
	15.5	45	8.8	53.0	142/123	150/130	6:15	766	817
2600	28.1	88	14.1	84.5	195/169	198/172	3:05	526	530
	24.4	75	12.3	74.0	183/159	186/161	4:20	682	695
	21.6	65	11.0	66.0	172/149	176/153	4:55	730	749
	18.8	55	9.8	59.0	159/138	164/142	5:35	773	802
	16.0	44	8.5	51.0	141/123	148/129	6:35	797	851
2400	28.0	79	12.6	75.5	187/162	190/165	3:30	569	575
	26.8	75	12.0	72.0	183/159	186/161	4:25	705	721
	23.6	65	10.7	64.0	172/149	176/153	5:05	758	782
	20.4	55	9.5	57.0	159/138	164/142	5:50	808	837
	16.5	43	8.1	48.5	138/120	147/128	7:00	834	895
2200	25.4	64	10.4	62.5	171/149	175/152	5:20	791	810
	22.2	55	9.2	55.0	159/138	164/142	6:05	834	863
	18.7	45	8.0	48.0	142/123	150/130	6:50	827	917
	17.5	42	7.7	46.0	136/118	126/109	7:20	864	930
2000	22.5	50	8.6	51.5	151/131	157/136	6:30	886	930
	20.5	45	8.0	48.0	142/123	150/130	7:20	897	917
	18.5	40	7.3	44.0	132/115	142/123	7:55	897	977

Table 10-16. Cruise Performance Using Best Power at 4,000 Feet.

MIXTURE SETTING:
1. Use FULL RICH mixture above 75% power. 2. Lean mixture in accordance with instructions in Section IV at 75% power and below.

RPM	MAN PRES (IN. HG)	% BHP	FUEL (GAL/HR)	FUEL (LBS/HR)	TRUE AIRSPEED MPH/KNOTS 2740 LBS	TRUE AIRSPEED MPH/KNOTS 2300 LBS	ENDURANCE (HR:MIN)	RANGE (NAUT MI) 2740 LBS	RANGE (NAUT MI) 2300 LBS
2700	26.5	87	14.2	85.0	197/171	200/174	3:10	521	539
	25.8	85	13.8	83.0	196/170	198/172	3:12	539	546
	23.2	75	12.5	75.0	186/161	189/164	4:10	673	692
	20.5	65	11.3	67.5	175/152	179/155	4:45	717	738
	17.9	55	10.0	60.0	161/140	166/144	5:25	752	782
	15.3	45	8.9	53.5	144/125	152/132	6:12	756	818
2600	26.2	82	13.3	79.5	193/168	196/170	3:20	556	566
	24.4	75	12.3	74.0	186/161	189/164	4:20	691	705
	21.5	65	10.9	65.5	175/152	179/155	4:52	734	756
	18.7	55	9.8	59.0	161/140	166/144	5:35	773	808
	15.7	44	8.5	51.0	123/107	130/113	6:30	799	847
2400	26.2	74	11.8	71.0	185/161	188/163	4:35	712	734
	23.3	65	10.7	64.0	175/152	179/155	5:05	756	787
	20.2	55	9.5	57.0	161/140	166/144	5:50	812	838
	16.7	44	8.2	49.0	143/124	150/130	6:50	843	892
2200	24.4	62	10.0	60.0	171/149	176/153	5:30	818	830
	22.0	55	9.2	55.0	161/140	166/144	6:00	838	871
	18.6	45	7.8	47.0	144/125	151/131	7:00	869	923
	17.6	42	7.7	46.0	142/123	147/128	7:20	871	934
2000	22.4	50	8.3	50.0	154/134	159/138	6:45	895	936
	20.4	45	7.8	46.5	144/125	151/131	7:15	908	963
	18.4	40	7.2	43.0	133/116	143/124	7:55	914	986

Table 10-17. Cruise Performance Using Best Power at 6,000 Feet.

MIXTURE SETTING:
1. Use FULL RICH mixture above 75% power. 2. Lean mixture in accordance with instructions in Section IV at 75% power and below.

RPM	MAN PRES (IN. HG)	% BHP	FUEL (GAL/HR)	FUEL (LBS/HR)	TRUE AIRSPEED MPH/KNOTS 2740 LBS	TRUE AIRSPEED MPH/KNOTS 2300 LBS	ENDURANCE (HR: MIN)	RANGE (NAUT MI) 2740 LBS	RANGE (NAUT MI) 2300 LBS
2700	24.7	81	13.3	80.0	195/169	200/174	3:20	573	573
	23.1	75	12.5	75.0	189/164	192/167	4:15	682	699
	20.4	65	11.3	67.5	178/155	181/157	4:45	725	747
	17.8	55	10.0	60.0	164/142	169/147	5:25	763	795
	15.2	45	8.9	53.5	145/126	155/134	6:15	782	834
2600	24.1	75	12.3	74.0	189/164	192/167	4:20	691	708
	21.3	65	10.9	65.5	178/154	181/157	4:55	743	765
	18.5	55	9.8	59.0	164/142	170/148	5:35	786	812
	15.8	45	8.6	51.5	145/126	155/135	6:25	810	864
2400	24.4	70	11.3	68.0	184/160	187/162	4:45	743	765
	22.8	65	10.7	64.0	178/154	181/157	5:05	769	795
	19.8	55	9.5	57.0	164/142	170/148	5:45	821	851
	16.5	44	8.0	48.0	143/124	153/133	6:50	847	906
2200	23.6	60	9.8	58.5	171/148	176/153	5:35	821	851
	21.8	55	9.2	55.0	164/142	170/148	6:00	847	882
	20.0	50	8.6	51.5	155/135	162/141	6:20	864	908
	17.6	43	7.8	46.5	141/123	151/131	7:10	878	943
2000	21.2	47	8.0	48.0	150/130	159/138	6:45	908	924
	19.2	42	7.4	44.5	139/121	150/130	7:20	912	986

Table 10-18. Cruise Performance Using Best Power at 8,000 feet.

MIXTURE SETTING:
1. Use FULL RICH mixture above 75% power. 2. Lean mixture in accordance with instructions in Section IV at 75% power and below.

RPM	MAN PRES (IN. HG)	% BHP	FUEL (GAL/HR)	FUEL (LBS/HR)	TRUE AIRSPEED MPH/KNOTS 2740 LBS	TRUE AIRSPEED MPH/KNOTS 2300 LBS	ENDURANCE (HR: MIN)	RANGE (NAUT MI) 2740 LBS	RANGE (NAUT MI) 2300 LBS
2700	23.6	75	12.6	75.5	195/169	197/171	4:10	686	704
	21.7	70	11.9	71.5	187/162	191/166	4:25	708	730
	20.4	65	11.3	67.5	181/157	185/161	4:40	730	756
	19.0	60	10.7	64.0	174/151	177/156	5:00	752	778
	17.8	55	10.1	60.5	167/145	173/150	5:20	769	801
	15.2	45	8.9	53.5	148/129	157/136	6:10	786	839
2600	23.0	71	11.8	71.0	189/164	193/168	4:20	721	738
	21.2	65	11.1	66.5	181/157	185/161	4:50	749	772
	19.8	60	10.4	62.5	174/151	179/156	5:10	773	795
	18.6	55	9.8	59.0	167/145	173/150	5:30	791	821
	17.0	50	9.3	55.5	158/137	165/143	5:50	808	843
	15.6	45	8.7	52.0	148/129	157/136	6:25	815	864
2400	22.8	64	10.6	63.5	181/157	185/161	5:10	784	808
	21.3	60	10.1	60.5	174/151	179/156	5:25	804	830
	19.8	55	9.5	57.0	167/145	173/150	5:45	825	856
	18.2	50	8.9	53.5	158/137	165/143	6:10	841	882
	16.4	44	8.2	49.0	146/127	155/135	6:45	847	908
2200	22.0	55	9.2	55.0	167/145	173/150	5:55	851	886
	20.0	50	8.6	51.5	158/137	165/143	6:25	873	917
	17.5	43	7.8	46.5	144/125	153/133	7:10	880	949
2000	20.3	45	7.8	46.5	148/129	156/136	7:10	917	975
	19.0	42	7.4	44.5	141/122	151/131	7:30	917	989

Table 10-19. Cruise Performance Using Best Power at 10,000 Feet.

MIXTURE SETTING:
1. Use FULL RICH mixture above 75% power. 2. Lean mixture in accordance with instructions in Section IV at 75% power and below.

RPM	MAN PRES (IN. HG)	% BHP	FUEL (GAL/HR)	FUEL (LBS/HR)	TRUE AIRSPEED MPH/KNOTS 2740 LBS	TRUE AIRSPEED MPH/KNOTS 2300 LBS	ENDURANCE (HR: MIN)	RANGE (NAUT MI) 2740 LBS	RANGE (NAUT MI) 2300 LBS
2700	21.4	70	11.9	71.5	190/165	195/169	4:25	715	740
	20.2	65	11.3	67.5	184/160	189/164	4:40	737	766
	18.8	60	10.7	64.0	178/155	183/159	5:00	760	793
	17.6	55	10.0	60.0	170/148	176/153	5:20	778	817
	16.3	50	9.5	57.0	161/140	169/147	5:40	792	836
	15.3	46	9.0	54.0	153/133	162/141	6:00	793	849
2600	21.1	65	11.0	66.0	184/160	189/164	4:50	756	782
	19.8	60	10.4	62.5	178/155	183/159	5:10	778	811
	18.3	55	9.8	59.0	170/148	176/153	5:25	798	838
	17.0	50	9.2	55.0	161/140	169/147	5:40	812	858
	15.5	46	8.8	52.5	153/133	162/141	6:05	819	876
2400	21.0	60	10.1	60.5	178/155	183/159	5:25	810	843
	19.5	55	9.5	57.0	170/148	176/153	5:45	832	871
	18.0	50	8.8	53.0	161/140	169/147	6:05	850	897
	16.8	46	8.4	50.5	153/133	162/141	6:30	858	912
2200	21.0	53	8.9	53.5	166/144	173/150	6:08	877	912
	19.8	50	8.6	51.5	162/141	169/147	6:12	884	928
	18.4	46	8.2	49.0	153/133	162/141	6:40	886	949
2000									

Table 10-20. Cruise Performance Using Best Power at 12,000 feet.

MIXTURE SETTING:
1. Use FULL RICH mixture above 75% power. 2. Lean mixture in accordance with instructions in Section IV at 75% power and below.

RPM	MAN PRES (IN. HG)	% BHP	FUEL (GAL/HR)	FUEL (LBS/HR)	TRUE AIRSPEED MPH/KNOTS 2740 LBS	TRUE AIRSPEED MPH/KNOTS 2300 LBS	ENDURANCE (HR: MIN)	RANGE (NAUT MI) 2740 LBS	RANGE (NAUT MI) 2300 LBS
2700	19.8	64	11.2	67.0	185/161	191/166	4:50	738	782
	18.8	60	10.7	64.0	180/156	186/162	5:00	756	798
	17.5	55	10.1	60.5	173/150	179/156	5:20	778	819
	16.2	50	9.5	57.0	165/143	171/149	5:40	791	843
	15.2	46	9.0	54.0	156/136	165/143	6:00	801	856
2600	19.6	60	10.4	62.5	180/156	186/162	5:10	778	817
	18.2	55	9.8	59.0	173/150	179/156	5:30	799	843
	16.8	50	9.3	55.5	165/143	171/149	5:50	815	864
	15.7	46	8.8	52.5	156/136	165/143	6:10	825	883
2400	19.6	56	9.6	57.5	174/151	180/156	5:40	825	899
	17.9	50	8.8	53.0	165/143	171/149	6:10	856	905
	16.4	45	8.3	49.5	154/134	162/141	6:35	863	926
2200	19.6	50	8.6	51.5	164/142	172/149	6:15	879	930
	18.0	45	8.0	48.0	154/131	162/141	6:50	893	960
2000									

Table 10-21. Cruise Performance Using Best Power at 14,000 Feet.

MIXTURE SETTING:
1. Use FULL RICH mixture above 75% power. 2. Lean mixture in accordance with instructions in Section IV at 75% power and below.

RPM	MAN PRES (IN. HG)	% BHP	FUEL (GAL/HR)	FUEL (LBS/HR)	TRUE AIRSPEED MPH/KNOTS 2740 LBS	TRUE AIRSPEED MPH/KNOTS 2300 LBS	ENDURANCE (HR: MIN)	RANGE (NAUT MI) 2740 LBS	RANGE (NAUT MI) 2300 LBS
2700	18.4	58	11.0	66.0	179/156	187/162	5:10	772	812
	17.4	55	10.1	60.5	174/151	182/158	5:15	782	825
	16.1	50	9.5	57.0	165/143	174/151	5:40	791	850
	15.6	48	9.3	55.5	160/139	171/149	5:50	792	860
2600	18.2	56	9.9	59.5	176/153	184/160	5:25	799	851
	16.7	50	9.2	55.0	165/143	174/151	5:50	812	873
	15.8	47	8.8	53.0	158/137	169/147	6:05	817	885
2400	18.2	52	9.1	54.5	169/147	177/154	5:55	847	902
	17.3	49	8.8	52.5	163/142	172/149	6:10	851	917
	16.4	46	8.4	50.5	156/136	167/145	6:30	853	930
2200									
2000									

Table 10-22. Landing Distances Using Normal Procedures.

ASSOCIATED CONDITIONS:

POWER--------------THROTTLE CLOSED
LANDING GEAR-------DOWN
WING FLAPS--------- FULL DOWN (33°)
WEIGHT--------------2740 LBS.

RUNWAY - PAVED, LEVEL, DRY SURFACE
APPROACH SPEED AT 50 FT - 81 MPH (71 KTS.) IAS

		PRESSURE ALTITUDE									
		Sea Level		2000 FT.		4000 FT.		6000 FT.		8000 FT.	
Wind Component Down Runway Knots	OAT °C	Ground Roll Feet	Total Over 50 Ft. Obstacle Feet	Ground Roll Feet	Total Over 50 Ft. Obstacle Feet	Ground Roll Feet	Total Over 50 Ft. Obstacle Feet	Ground Roll Feet	Total Over 50 Ft Obstacle Feet	Ground Roll Feet	Total Over 50 Ft Obstacle Feet
	-20	773	1805	904	1911	1046	2103	1193	2373	1343	2667
	-10	804	1851	940	1969	1087	2107	1240	2450	1401	2755
	0	834	1906	976	2028	1129	2238	1287	2526	1454	2842
0	10	865	1962	1011	2089	1170	2305	1334	2603	1508	2930
	20	896	2018	1047	2149	1211	2372	1382	2680	1561	3017
	30	926	2074	1083	2209	1253	2439	1429	2757	1614	3105
	40	956	2129	1118	2269	1294	2507	1476	2834	1667	3193
	-20	728	1700	854	1805	990	1990	1131	2250	1281	2534
	-10	758	1746	889	1862	1030	2056	1177	2326	1332	2620
	0	788	1800	924	1920	1070	2122	1223	2400	1384	2706
10	10	818	1855	958	1979	1110	2187	1268	2475	1436	2791
	20	848	1910	993	2038	1150	2253	1315	2551	1488	2877
	30	877	1964	1028	2097	1191	2319	1361	2626	1540	2963
	40	906	2018	1062	2155	1231	2386	1407	2702	1592	3036
	-20	688	1601	808	1703	938	1882	1074	2132	1218	2406
	-10	717	1646	841	1758	977	1946	1119	2206	1269	2491
	0	745	1698	875	1814	1017	2012	1163	2279	1319	2574
20	10	774	1751	909	1874	1055	2075	1208	2353	1370	2658
	20	804	1806	943	1931	1094	2139	1253	2426	1421	2742
	30	832	1859	977	1989	1134	2204	1298	2501	1471	2827
	40	860	1911	1010	2046	1173	2269	1343	2575	1522	2884

NOTE: Maximum demonstrated crosswind velocity is 12 MPH (11 Knots).

Table 10-23. Maximum Performance Landing Distances.

POWER---------------- THROTTLE CLOSED
LANDING GEAR-------- DOWN
WING FLAPS----------- FULL DOWN (33°)
WEIGHT----------------2740 LBS.

RUNWAY - PAVED. LEVEL, DRY SURFACE
APPROACH SPEED AT 50 FT. - 75 MPH (65 KTS) IAS

		PRESSURE ALTITUDE									
		Sea Level		2000 FT.		4000 FT.		6000 FT.		8000 FT.	
Wind Component Down Runway Knots	OAT °C	Ground Roll Feet	Total Over 50 Ft. Obstacle Feet	Ground Roll Feet	Total Over 50 Ft. Obstacle Feet	Ground Roll Feet	Total Over 50 Ft. Obstacle Feet	Ground Roll Feet	Total Over 50 Ft. Obstacle Feet	Ground Roll Feet	Total Over 50 Ft. Obstacle Feet
	-20	676	1457	735	1585	792	1744	854	1958	911	2219
	-10	703	1501	764	1633	823	1798	888	2019	947	2290
0	0	730	1545	793	1681	855	1852	921	2081	983	2361
	10	757	1588	622	1730	886	1906	955	2142	1019	2432
	20	783	1632	851	1778	917	1960	989	2204	1055	2503
	30	810	1675	880	1826	949	2014	1023	2265	1091	2574
	40	837	1719	909	1875	980	2068	1056	2327	1127	2644
	-20	642	1375	699	1499	755	1653	816	1860	871	2110
	-10	668	1418	727	1546	785	1706	849	1919	907	2181
10	0	695	1462	750	1594	816	1758	881	1980	942	2250
	10	721	1504	789	1641	846	1811	914	2040	977	2320
	20	746	1547	813	1689	877	1865	947	2100	1012	2389
	30	773	1590	841	1736	908	1918	981	2161	1048	2460
	40	799	1632	869	1784	939	1971	1013	2221	1083	2528
	-20	611	1297	667	1417	721	1565	780	1763	835	2005
	-10	636	1338	694	1463	750	1616	812	1822	869	2073
20	0	662	1381	722	1509	781	1669	844	1882	903	2141
	10	688	1423	749	1556	810	1720	876	1940	938	2210
	20	713	1466	777	1602	840	1773	909	2000	972	2278
	30	738	1506	805	1649	870	1824	941	2059	1007	2347
	40	764	1549	832	1696	900	1877	973	2119	1041	2414

NOTE: Maximum demonstrated crosswind velocity is 12 MPH (11 Knots).

Table 10-24. Cruise Power Schedule.

1. BEST POWER IS 125° RICH OF PEAK T.I.T.
2. BEST ECONOMY IS PEAK T.I.T. (NOT TO EXCEED 1650°F)

BEST POWER
(BEST ECONOMY - ADD 0.8" TO MP LISTED)

Pressure Altitude Feet / Std. Day	Fuel Flow / Standard Temperature	75% POWER (157.5 BHP)						70% POWER (147 BHP)						65% POWER (136.5 BHP)					
	RPM	2200	2300	2400	2500	2600	2700	2200	2300	2400	2500	2600	2700	2200	2300	2400	2500	2600	2700
	Best Economy		10.9	11.1	11.2	11.3	11.3	10.2	10.4	10.6	10.6	10.6	10.6	9.6	9.8	10.0	10.1	10.1	10.1
	Best Power		13.0	13.1	13.3	13.3	13.3	12.1	12.3	12.4	12.5	12.5	12.5	11.3	11.5	11.7	11.8	11.9	11.9
	Standard Temperature	MANIFOLD PRESSURE - INCHES OF MERCURY																	
S.L.	15°C		34.6	33.5	32.5	31.5	30.3	33.7	32.6	31.7	30.7	29.8	28.8	31.8	30.7	29.8	28.9	28.0	27.1
2000	11°		34.1	33.1	32.2	31.2	30.1	33.3	32.2	31.3	30.4	29.5	28.5	31.4	30.4	29.4	28.5	27.7	26.8
4000	7°		33.9	32.8	31.8	30.8	29.8	32.8	31.9	30.9	30.0	29.1	28.0	30.9	29.9	28.9	28.1	27.3	26.4
6000	3°		33.3	32.4	31.5	30.5	29.5	32.5	31.4	30.5	29.6	28.7	27.9	30.4	29.4	28.5	27.7	27.0	26.1
8000	-1°		33.2	32.3	31.3	30.4	29.4		31.3	30.3	29.5	28.5	27.7	30.3	29.3	28.3	27.5	26.7	25.9
10000	-5°			32.0	30.9	30.1	29.1		31.1	29.9	29.0	28.1	27.3	29.8	28.9	27.8	27.1	26.3	25.5
12000	-9°			31.7	30.7	29.8	29.0			29.6	28.8	27.9	27.1		28.7	27.6	26.9	26.1	25.3
14000	-13°				30.4	29.6	28.6			29.3	28.4	27.7	26.9			27.3	26.5	25.8	25.0
16000	-17°				30.3	29.4	28.5				28.3	27.5	26.7			27.1	26.3	25.6	24.7
18000	-21°					29.3	28.5				28.1	27.3	26.5				26.2	25.5	24.6
20000	-25°					29.3	28.5					27.5	26.6					25.5	24.7
22000	-29°						28.5					27.6	26.5					25.6	24.6
24000	-33°						28.5						26.6						24.6

EXAMPLE:
CRUISE ALT. 8500' PRESS.
OAT: 10°
POWER: 75%
RPM: 2500
M.P. BEST POWER: 32.4" Hg (12° corr)
M.P. BEST ECONOMY: 33.2" Hg (add 0.8")

NOTE: 1. ADD 1" M.P. FOR EACH 10°C OAT ABOVE STANDARD DAY TEMPERATURE. SUBTRACT 1" M.P. FOR EACH 10° OAT BELOW STANDARD DAY TEMPERATURE. IF OAT ABOVE STANDARD PRECLUDES OBTAINING THE DESIRED M.P., USE THE NEXT HIGHER RPM/M.P. WITH APPROPRIATE TEMPERATURE CORRECTION TO M.P.

2. INTERMITTENT ROUGHNESS MAY OCCUR ABOVE 20.000' BELOW 2500 RPM: AND BELOW 16" M.P. AT ALL ALTITUDES.

DO NOT EXCEED 1650° T.I.T.

Table 10-25. Cruise Power Schedule.

1. BEST POWER IS 125° RICH OF PEAK T.I.T.
2. BEST ECONOMY IS PEAK T.I.T. (NOT TO EXCEED 1650°F)

BEST POWER
(BEST ECONOMY - ADD 0.8" TO MP LISTED)

Pressure Altitude Feet Std. Day	Fuel Flow / Standard Temperature	55% POWER (115.5 BHP)						45% POWER (94.5) BHP)						35% POWER (73.5 BHP)					
	RPM	2200	2300	2400	2500	2600	2700	2200	2300	2400	2500	2600	2700	2200	2300	2400	2500	2600	2700
	Best Economy	8.2	8.4	8.6	8.7	8.8	8.8	6.9	7.1	7.3	7.4	7.5	7.6	5.7	5.8	6.1	6.3	6.5	6.6
	Best Power	9.8	10.0	10.3	10.5	10.6	10.7	8.4	8.6	8.9	9.2	9.4	9.5	6.7	7.2	7.6	7.9	8.1	8.3
	Standard Temperature	MANIFOLD PRESSURE - INCHES OF MERCURY																	
S.L.	15°C	28.0	27.1	26.3	25.5	24.7	24.0	24.5	23.6	22.8	22.1	21.5	20.8	20.9	20.1	19.5	18.9	18.3	17.7
2000	11°	27.6	26.7	25.8	25.1	24.4	23.7	24.1	23.2	22.5	21.7	21.2	20.5	20.4	19.6	19.1	18.6	17.9	17.4
4000	7°	27.1	26.2	25.3	24.6	23.9	23.1	23.5	22.6	21.9	21.3	20.6	20.0	20.0	19.3	18.5	18.0	17.5	17.0
6000	3°	26.8	25.8	24.8	24.2	23.5	22.7	23.0	22.2	21.4	20.8	20.2	19.5	19.4	18.7	18.0	17.5	16.8	16.4
8000	-1°	26.3	25.4	24.5	23.8	23.2	22.4	22.6	21.7	21.0	20.4	19.7	19.1	19.0	18.2	17.5	16.8	16.2	15.7
10000	-5°	25.9	25.0	24.1	23.5	22.7	22.0	22.1	21.2	20.6	19.9	19.4	18.6	18.5	17.8	17.1	16.5	15.9	
12000	-9°	25.5	24.7	23.8	23.2	22.5	21.8	21.7	21.0	20.3	19.6	19.0	18.4	18.1	17.4	16.7	16.0		
14000	-13°		24.3	23.5	22.8	22.2	21.4	21.3	20.6	20.0	19.4	18.6	17.9	17.6	17.1	16.5	15.8		
16000	-17°			23.3	22.6	22.0	21.2	21.0	20.4	19.7	19.0	18.4	17.7	17.4	16.8	16.3			
18000	-21°			23.2	22.4	21.6	21.0		20.1	19.5	18.8	18.1	17.4	17.2	16.5	15.9			
20000	-25°				22.6	21.8	21.1			19.5	18.8	18.1	17.4						
22000	-29°				22.7	21.8	21.0				18.8	18.1	17.4						
24000	-33°					22.1	21.1				19.4	18.4	17.4						

NOTE:

1. ADD 1" M.P. FOR EACH 10°C OAT ABOVE STANDARD DAY TEMPERATURE. SUBTRACT 1" M.P. FOR EACH 10°C OAT BELOW STANDARD DAY TEMPERATURE. IF OAT ABOVE STANDARD PRECLUDES OBTAINING THE DESIRED M.P., USE THE NEXT HIGHER RPM/M.P. WITH APPROPRIATE TEMPERATURE CORRECTION TO M.P.
2. INTERMITTENT ROUGHNESS MAY OCCUR ABOVE 20,000' BELOW 2500 RPM; AND BELOW 16" M.P. AT ALL ALTITUDES.

DO NOT EXCEED 1650° T.I.T.

Table 10-26. Power Schedule.

RECOMMENDED LONG RANGE CRUISE
Best Economy
(Leaned To Peak TIT)

Std. Day Altitude / % BHP	RPM (MP-"Hg.)									
	75%	70%	65%	60%	55%	50%	45%	40%	35%	Std. Temp.
S. L.	2300 (35.4)	2250 (33.5)	2100 (33.4)	2050 (32.4)	2000 (30.8)	2000 (28.9)	2000 (27.1)	2000 (25.3)	2000 (23.4)	15°
2000	2300 (35.0)	2250 (33.4)	2100 (33.0)	2050 (31.6)	2000 (30.1)	2000 (28.3)	2000 (26.5)	2000 (24.7)	2000 (22.8)	11°
4000	2300 (34.6)	2250 (33.2)	2100 (32.7)	2050 (31.2)	2000 (29.7)	2000 (27.8)	2000 (26.0)	2000 (24.2)	2000 (22.5)	7°
6000	2350 (33.9)	2250 (32.7)	2150 (31.8)	2100 (30.3)	2000 (29.4)	2000 (27.5)	2000 (25.5)	2000 (23.9)	2000 (22.0)	3°
8000	2350 (33.8)	2300 (32.2)	2250 (30.6)	2150 (28.7)	2100 (28.2)	2000 (27.2)	2000 (25.2)	2000 (23.4)	2000 (21.5)	-1°
10000	2400 (32.8)	2350 (31.3)	2300 (29.8)	2200 (28.7)	2150 (27.3)	2100 (25.8)	2000 (25.0)	2000 (23.1)	2000 (21.1)	-5°
12000	2450 (32.0)	2400 (30.5)	2350 (29.1)	2300 (27.4)	2250 (25.9)	2150 (25.0)	2100 (23.6)	2000 (22.5)	2000 (20.7)	-9°
14000	2500 (31.2)	2450 (29.7)	2400 (28.1)	2350 (26.7)	2300 (25.1)	2250 (23.7)	2200 (22.1)	2150 (20.8)	2050 (20.0)	-13°
16000	2550 (30.6)	2500 (29.3)	2450 (27.5)	2400 (26.0)	2350 (24.5)	2300 (23.1)	2250 (21.6)	2200 (20.0)	2150 (18.5)	-17°
18000	2600 (30.2)	2550 (28.6)	2500 (27.0)	2450 (25.6)	2400 (24.1)	2400 (22.3)	2350 (20.7)	2300 (19.2)	2250 (17.7)	-21°
20000	2650 (29.8)	2600 (28.[illegible])	2550 (26.9)	2500 (25.4)	2500 (23.4)	2450 (22.0)	2400 (20.4)	2100 (18.5)	ROUGH OPERATION	-25°
22000	2700 (29.1)	2650 (27.6)	2600 (26.5)	2600 (24.6)	2550 (23.1)	2500 (21.5)	2500 (19.8)	2500 (17.7)		-29°
24000	2700 (28.0)	2700 (27.4)	2700 (25.6)	2650 (24.2)	2650 (22.4)	2600 (21.2)	2550 (19.8)	2550 (17.8)		-33°

For Fuel Flow See Mi/Gal Curves

CONDITIONS:

1. Continuous operation at TIT in excess of 1650°F is prohibited.
2. **Manual leaning to 1550°F TIT permitted as long as CHT remains below 425° and oil temp. below 220°, excluding climb.**
3. Recommended Max. Cruise CHT is 440°F and oil temp. is 220°F.
4. Continuous operation at mixtures leaner than peak TIT prohibited.

NOTE: Correct MAP 1" for each 10°C variation from standard.

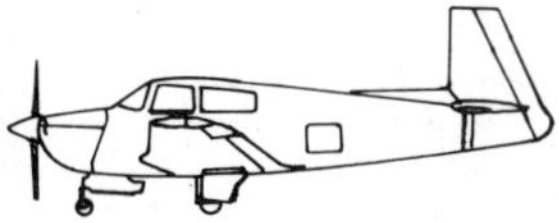

Index